AF531616

Biological Control of Pest and Weed Management

About the Authors

Dr. Awanindra Kumar Tiwari is currently working as a Scientist- Plant Protection (Entomology) at Krishi Vigyan Kendra, Raebareli of Chandra Shekhar Azad University of Agriculture and Technology, Kanpur, UP. He has completed his Ph.D. (Agricultural Zoology and Entomology) from the Department of Zoology, Faculty of Science, University of Allahabad, Prayagraj. He has 15 years' experience as a Scientist. He obtained the Excellence in Extension Award and the Best KVK Scientist Award. He has published many national and international research papers, many Books, chapters, and popular articles. He has participated in and presented papers on different topics at national and international workshops, seminars, and conferences. Dr. Tiwari has vast experience and expertise in Apiculture, Economic Entomology, Applied Entomology, Biological Control, Integrated Pest Management, Organic Farming, and Natural Farming.

Dr Wajid Hasan, Ph.D. PDF, is devoted to research and extension activities in Agriculture Entomology. Dr Hasan is a Subject Matter Specialist in Entomology at KVK Jehanabad, Bihar Agricultural University in Bihar. He was awarded Dr DS Kothari UGC-Post Doctoral Fellow (PDF) in 2010, offered by the University Grant Commission of India. He holds a doctorate in Agricultural Entomology from GB Pant University of Agri & Tech, Pantnagar, UK, India. He has good terms with farmers, and he serves them as an adviser for plant protection measures. Dr Hasan has received eleven awards from several societies for his outstanding contribution to the relevant discipline. He has 75 research papers, 42 books, 65 book chapters, 45 popular articles, 140 success stories, 3 Practical Manuals, 16 success stories and 52 delivered lectures in different seminars/symposia/conferences/workshops and seven radio talks. Dr Hasan stayed as the chief organizer for eleven international conferences with more than three thousand participants each and managed a team of about three hundred co-organizers each. He has also organized 589 training programs for farmers, rural youth and agricultural professionals with 24250 participants. In addition, he has conducted three skill development training programs of 200 hours each and 45 On-Farm Trials (OFT). Dr Hasan has a lifetime membership in 6 scientific societies. He is the Editor-in-Chief of the International Journal of Agricultural and Applied Sciences (IJAAS) and an Editorial board member for 5 scientific International Journals.

Dr. Milind D. Joshi, Subject Matter Specialist (Plant Protection), Agricultural Development Trust's Krishi Vigyan Kendra, Baramati, has more than 15 years of experience in research and extension. He has a Ph.D. in Agricultural Entomology. He has been awarded 8 various state & national-level awards for his contribution to the field of agriculture. He has visited 6 countries viz. The Netherlands, Turkey, Belgium, France, Spain & Indonesia for exposure. He is also the Editorial Member for various journals and magazines. He is also the Technical Committee Member and Advisor for various organizations. He has attended more than 31 seminars, conferences, and workshops and about 19 trainings, summer schools, and winter schools. He has published over 35 professional research papers in International and National refereed Journals in agriculture. He also has more than 50 popular articles and about 30 radio talks for the benefit of the farming community. He is the Principal Investigator for the ICAR-funded Project on All India Coordinated Research Project on Honey Bee & Pollinators for Maharashtra State. He also has been the Principal Investigator for DBT and other projects of government & private funding.

Dr. Pradeep Natthuji Dawane is engaged in research at, the Department of Agriculture Entomology, Regional Fruit Research station Katol, Dr. Panjabrao Deshmukh Krishi Vidyapeeth, Akola, Maharashtra. He completed his Ph.D. (Agril. Entomology) in 2013 from Dr. Panjabrao Deshmukh Krishi Vidyapeeth, Akola, (M.S.), M. Sc. (Agril. Entomology) in 1998 from College of Agriculture, Dr. Panjabrao Deshmukh Krishi Vidyapeeth, Akola, (M. S.) with remarkable distinction of being in his field of study. He has attended several National and International conferences, Seminars, and Conclave, where he received several prestigious Awards for his research including the Best Oral Presentation Award. He has been Awarded the Distinguished Scientist Award 2023 and Eminent Scientist Award 2023. The author has published more than 25 Research Articles and 5 Popular Articles in National and International journals and Magazines. One textbook published Introductory Nematology by Agromet publishers Dharmpeth, Nagpur (M. S.).

Dr Radhika Negi is presently working as a Subject Matter Specialist (Vegetable Science and Floriculture) at CSK HPKV, KVK, Lahaul, and Spiti 1 at Kukumseri. She has done her post-graduation in Vegetable Science from Dr YSP UHF, Nauni, Solan, and qualified ASRB-ICAR NET in Vegetable Science in 2018. She has 3 years and 8 months of experience in the field of extension and research and is involved in various projects. In recognition of their outstanding contribution to the field of Vegetable Science and Floriculture, she was honored with the Young Scientist Award and begged best paper presentation certificate in 3 days International Conference on "Advances in Agriculture, Veterinary and Allied Science for Improving Livelihood and Environmental Security-2022". She published 7 research papers and review papers/articles in international and national reputed journals, 6 book chapters, 8 popular articles, and 5 abstracts published in different conferences. Besides this, she has conducted and attended numerous training programs (on and off campus), workshops and seminars, etc

Dr. Sheetanshu Gupta working in Biochemistry, Molecular Biology, and Biotechnology for the last 12 years, an alumnus of G.B. Pant University of Agriculture & Technology, Pantnagar, Uttarakhand, India, ICAR ARS NET (Plant Biochemistry and Plant Physiology) and CSIR NET. Dr. Gupta gained valuable experience in the areas of environmental problems, aeroponics and hydroponics, and protein biochemistry. His work showcased his expertise in applying innovative farming techniques and exploring novel approaches to plant protection. Dr. Gupta researched tissue culture hydroponics and mushroom farming, contributing to the advancement of these fields. Dr. Gupta's academic qualifications, research experience, and expertise in biochemistry and biotechnology make him a valuable asset in the field. With his passion for teaching and research, dedication to innovative farming techniques, and student advocacy, Dr. Gupta continues to contribute significantly to the advancement of agricultural and biochemical sciences with more than 10 published books and nearly 20 Research/ Review papers

Dr. Dhirendra Kumar is a distinguished Scholar, with more than 15 years of teaching and industrial experience. Currently working as an Assistant Professor in the Department of Botany at Chaudhary Bansi Lal University, Haryana, India. His areas of interest are Plant Biology, Biotechnology, Bioprocess Development, Termite Biology, and Waste Management and currently working on the green synthesis of nanomaterials. In addition, he also received the Outstanding Achievement Award 2021 at the 3rd International Conference (GIAFAS-2021), the Best Young Faculty Award GAREA-2021(Global Academic Research Excellence Award) India, Young Botanist Award 2020" from the Deccan Environmental Research Organization, Bijapur, Karnataka, India. Young Scientist Awards 2015 and 2022 Sponsored by CGCOST (Chhattisgarh Council of Science and Technology), Chhattisgarh State, GAFEF with Tribhuvan University Nepal and India. Currently, he has completed one University research project, has five granted patents (One Australian and four Indian), Four books, 10 book chapters, and more than 25 research papers in good impact International and National reputed Journals with 20+ national and international Conferences/ Seminars paper and oral presentations in India and neighboring countries

Biological Control of Pest and Weed Management

Awanindra Kumar Tiwari
Krishi Vigyan Kendra, Raebareli of Chandra Shekhar Azad University of Agriculture and Technology, Kanpur, U.P., India

Wajid Hasan
Senior Scientist, Krishi Vigyan Kendra
Jehanabad of Bihar Agricultural University, Bihar, India

Milind D. Joshi
Agricultural Development Trust's Krishi Vigyan Kendra
Baramati, Pune, Maharashtra, India

Pradeep Natthuji Dawane
Department of Agriculture Entomology
RFRS, Katol, Dr. PDKV, Akola, Maharashtra, India

Radhika Negi
Vegetable Science and Floriculture
CSK HPKV, KVK, Lahaul, and Spiti 1 at Kukumseri, India

Sheetanshu Gupta
Department of Biochemistry, G.B. Pant
University of Agriculture & Technology, Pantnagar, Uttarakhand, India

Dhirendra Kumar
Department of Botany, Chaudhary Bansi Lal University
Haryana, India

1168, Sector 13, Urban Estate
Karnal-132 001, Haryana
Tel: 91-84470 75807, 18440 41168
Email: contentvibesppa@gmail.com
www.contentvibes.in

Print ISBN: 978-81-97719-28-8

ebook-ISBN: 978-81-97719-24-0

Perface

Biological control stands as a pivotal strategy for managing pest populations through environmentally sustainable methods. As the detrimental impacts of conventional pesticides continue to surface, a growing consensus among scientists, policymakers, and practitioners supports the need to embrace more ecological pest management methods. "Biological Control of Insect Pests and Weeds" is a comprehensive textbook designed to meet this demand, offering detailed insights into the use of natural processes for controlling pests and weeds.

This book is aimed at providing a thorough grounding in the principles, history, and practical applications of biological control. It targets a diverse audience, including students, researchers, and professionals in entomology, ecology, and agricultural sciences, who seek to deepen their knowledge and apply biological control principles effectively. Bridging the gap between theoretical research and practical applications, this textbook serves both as an academic resource and a practical guide.

The concept of biological control involves using living organisms—such as insects, mites, nematodes, and microorganisms—to regulate the population of pest species naturally and sustainably. This approach helps reduce the dependence on chemical pesticides, which can lead to environmental degradation and resistance among pest populations, while also promoting biodiversity and ecological balance. Given the interdisciplinary nature of biological control, this textbook provides a comprehensive examination of the biological, ecological, and technological aspects involved.

"Biological Control of Insect Pests and Weeds" unfolds the narrative of biological control through a detailed exploration of its historical background, foundational principles, and the contemporary advancements that shape its application today. The textbook begins with a historical overview, tracing the evolution of biological control practices from ancient times to present-day scientific methodologies, highlighting significant milestones and key figures.

The book explains the ecological rationale behind using natural enemies to manage pest populations, discussing dynamics of predator-prey relationships, host-parasite interactions, and the role of microbial pathogens in natural pest suppression. This section lays the groundwork for understanding how biological control integrates with broader ecological systems and agricultural practices.

Dedicated chapters discuss various agents of biological control—parasitoids, predators, pathogens, and weed-eating species. Each group is examined in detail, with specific focus on their biology, behavior, and roles in pest management. This includes a thorough review of the mechanisms through which these agents target pests, strategies for enhancing their effectiveness, and the practical challenges involved.

Significant attention is given to the methodologies involved in implementing biological control, including the breeding, mass production, and field release of biocontrol agents. Step-by-step guides detail the processes from laboratory rearing to field evaluation, supplemented by case studies that illustrate successful implementations. Economic and logistical considerations are also explored, offering a comprehensive view of the planning and execution of biological control strategies.

Moreover, the textbook addresses advanced technologies and innovations revolutionizing biological control, such as genetic engineering, the use of semiochemicals, and the integration of biotechnological tools, providing insights into how modern science is enhancing the efficacy and scope of biological control methods.

The final chapters reflect on the future of biological control, discussing ongoing research, potential technological advancements, and ethical and regulatory considerations. By providing a forward-looking perspective, the textbook encourages readers to consider the evolving challenges and opportunities in biological control.

In summary, "Biological Control of Insect Pests and Weeds" is not merely a textbook—it is a guide to sustainable agricultural practices and a testament to the power of ecological science. It equips its readers with the knowledge and tools to advocate for, implement, and advance biological control methods, contributing to a more sustainable and environmentally responsible future for global agriculture

Contents

Unit I

History, principles and scope of biological control; important groups of parasitoids, predators and pathogens; principles of classical biological control-importation, augmentation and conservation. History of insect pathology, infection of insects by bacteria, fungi, viruses, protozoa, rickettsiae, spiroplasma and nematodes. Plant Protection–Entomology

A. History, Principles, and Scope of Biological Control

Biological Control

Biological control, defined as the use of living organisms to reduce the population of pest species, stands out as one of the most successful, cost-effective, and environmentally friendly methods of pest management. This natural process is intrinsic to maintaining ecological balance and supporting the greenery of our planet by enabling plants to produce the necessary biomass to sustain diverse forms of life.

The Essence and Impact of Biological Control

Biological control is an omnipresent force in both natural and human-made ecosystems, constantly active without any human intervention. Its origins trace back over 500 million years with the evolution of the first ecosystems. The process involves naturally occurring reductions in pest populations, which historically, have allowed for the flourishing of plant life critical to the earth's ecological systems. Without these biological control mechanisms, the amount of energy produced by plants would be a fraction of current levels, drastically reducing the earth's capacity to support life.

The history of utilizing biological control dates back to around 300 AD when humans first used predatory ants to manage pests in citrus orchards. A significant milestone in the field was reached in 1888 with the introduction of *Rodolia cardinalis*, a species of ladybird beetle, to combat scale insects in California's citrus groves. This event marked the beginning of the large-scale use of biological control, which has since secured numerous successes, yielding substantial economic benefits through sustainable pest management practices.

Necessity and Future of Biological Control

The urgency for advanced biological control methods is underscored by several pressing global challenges:

1. The growing need to feed an estimated 11 billion people by the end of the century.
2. The depletion of fossil fuels, underscores the diminishing availability of conventional synthetic pesticides.
3. The critical need to reduce environmental pollution and conserve biodiversity after decades of decline.

These factors compel a shift in agricultural research towards a systems approach where pest management becomes a central component, influencing all aspects of agricultural practices from crop system design to harvest. The future of pest management is expected to rely heavily on biological control due to its sustainability, cost-effectiveness, and minimal environmental impact.

Economic and Ecological Benefits

Biological control offers significant economic advantages, with classical biological control alone applied over 350 million hectares, yielding benefit-cost ratios between 20:1 and 500:1. Moreover, natural biological control, which operates continuously across all terrestrial ecosystems, is estimated to provide services worth at least $400 billion annually— a stark contrast to the $8.5 billion spent annually on insecticides.

Augmentative biological control, though used on a smaller scale (covering 0.4% of cultivated land), demonstrates benefit-cost ratios comparable to or better than those of chemical pest control, underscoring its economic viability and efficiency.

Widespread Adoption and Success

Over the past 120 years, more than 5,000 introductions of approximately 2,000 species of exotic arthropod agents have been implemented across 196 countries or islands. Currently, over 150 species of natural enemies, including parasitoids, predators, and pathogens, are commercially available, illustrating the widespread adoption and success of biological control strategies globally.

1. History of Biological Control

Biological control, as a formal scientific discipline, began in the late 19th century, but its practices can be traced back to ancient times. Historical records from China around 300 AD mention the use of predatory ants to control pests in citrus orchards. However, the modern era of biological control commenced with the successful introduction of the *Rodolia cardinalis* (vedalia beetle) to California in 1888 to combat the cottony cushion scale, a major threat to the citrus industry. This event marked a significant milestone, demonstrating the potential of using natural enemies to manage pest populations and setting the precedent for future biological control programs.

Historical Milestones in Biological Control

1. Early Beginnings and Conceptual Foundations

Biological control, or the use of natural enemies to manage pest populations, has ancient roots. Historical documentation suggests that as early as 300 AD, Chinese farmers used predatory ants to protect their citrus groves from caterpillars and large boring beetles. This practice indicates an early understanding of biological interactions and their potential for pest control. In Europe, during the Middle Ages, cultural controls and the encouragement of natural predators were recognized methods for protecting crops.

However, the systematic scientific exploration and application of biological control began much later. The landmark event in the modern era of biological control occurred in 1888, with the introduction of the *Rodolia cardinalis*, commonly known as the vedalia beetle, into California from Australia to control the cottony cushion scale. This pest was devastating to the California citrus industry, and the successful mitigation through a biological control agent was a turning point that showcased the effectiveness of this approach on a large scale.

2. Establishment of Research Agencies and Expansion

The success of the Vedalia Beetle led to the establishment of the first government research agencies focused on biological control. In the early 1900s, countries such as the United States began investing in biological control research, forming dedicated units within agricultural departments. These agencies aimed to identify, research, and implement biological control solutions across various agricultural sectors.

By the mid-20th century, biological control had gained a solid footing within the scientific community. The 1960s and 1970s saw a significant expansion of biological control strategies into integrated pest management (IPM) programs. This period was marked by increased ecological awareness and the recognition of the detrimental impacts of synthetic pesticides, which spurred research into more sustainable pest management options. Biological control was seen as a cornerstone of IPM strategies, integrating chemical, cultural, and biological methods to manage pest populations in an environmentally and economically sustainable way.

3. Global Spread and Institutional Support

The latter half of the 20th century witnessed the global spread of biological control practices. International cooperation, such as the work done by the International Organization for Biological Control (IOBC), facilitated the exchange of knowledge and techniques among countries. Numerous international projects were launched, especially in developing countries where agricultural pests posed significant challenges to food security and economic development.

Institutions such as the Commonwealth Agricultural Bureaux International (CABI) and the Food and Agriculture Organization (FAO) of the United Nations played crucial roles in promoting and supporting biological control projects around the world. These organizations helped standardize methodologies and bolster the scientific and regulatory frameworks necessary for the successful implementation of biological control.

4. Technological Advances and Modern Practices

The late 20th and early 21st centuries have seen remarkable technological advances that have further refined and expanded the scope of biological control. Genetic engineering and biotechnological innovations have opened new avenues for enhancing the efficacy and specificity of biological control agents. For example, the development of genetically modified viruses that are more effective and selective in targeting pest species has been a significant advancement.

Additionally, the advent of molecular biology tools has improved the understanding of the interactions between pests and their natural enemies. These tools help in the precise identification and monitoring of biological control agents, ensuring their effectiveness and safety when released into new environments.

5. Current Trends and Future Perspectives

Today, biological control is recognized as a vital component of sustainable agricultural practices and environmental conservation. The field continues to evolve with advancements in science and technology. Current trends include the use of information technology and artificial intelligence to model pest and predator interactions, predict outbreaks, and optimize the release and management of biological control agents.

Looking forward, the integration of biological control strategies with other technological innovations in agriculture, such as precision farming and automated monitoring systems, promises to enhance the efficiency and impact of biological control programs. As global awareness of environmental sustainability grows, the importance of biological control in maintaining ecological balance and supporting sustainable agriculture is increasingly acknowledged.

Table: Highlights in Entomology and Discovery of Parasitoids

S No.	Year	Discoverer	Contribution	Details
1	**ca. 310**	Aristoteles (Greece, 384 - 322 BC)	Historia Animalum	Documented natural history and taxonomy of animals, foundational work in zoology including some observations on insect life.
2	**ca. 300**	Guo Pu (China, 276 - 324)	Commentary on the Literary Expositor	Mentioned tachinid parasitoid but lacked understanding of its biology.
3	**1096**	Lu Dian (China, 1042 - 1102)	New Additions to the Literary Expositor	Described the full cycle of insect parasitism by a tachinid parasitoid; recognized as the first detailed description of insect parasitism based on complete lifecycle observations.
4	**1321**	Dante Alighieri (Italy, 1265 - 1321)	La Divina Commedia	Included many references to insects within this literary epic.
5	**1551-1634**	Conrad Gessner (Germany, 1516 - 1565)	Historia Animalum	An encyclopedic work summarizing earlier information along with his observations, classification of animals including insects; the volume on insects published posthumously.
6	**1552**	Edward Wotton (Britain, 1492 - 1555)	De Differentiis Animalium Libri Decem	Encyclopedic work summarizing earlier information on animal differences, including some insect descriptions.
7	**1602**	Ulisse Aldrovandi (Italy, 1522 - 1605)	De Animalibus Insectis Libri VII	Documented emergence of parasitoid larvae from caterpillar; did not fully understand the phenomenon; considered the first work dedicated to entomology.
8	**1660**	John Ray (Britain, 1627 - 1705)	Catalogus Plantarum circa Cantabrigiam nascentium	Observed emergence of parasitoid larvae from caterpillar in 1658.
9	**1662**	Johannes Goedaert (Holland, 1617-1668)	Metamorphosis Naturalis	Illustrated and described emergence of larvae and adults of parasitoids across three volumes, without understanding parasitism fully.

10	**1668**	Francesco Redi (Italy, 1626 - 1697)	Esperienze Intorno alla Generazione degli Insetti	Documented emergence of parasitoid larvae from hosts but did not understand the phenomenon of parasitism.
11	**1669**	Jan Swammerdam (Holland, 1637 - 1680)	Historia Insectorum Generalis	Observed many parasitoids in larval, pupal, and adult stages; categorized internal/external parasitoids, but did not observe oviposition; provided the first correct European interpretation of insect parasitism.
12	**1670/71**	Martin Lister (Britain, 1639 - 1712)	Philosophical Transactions of the Royal Society	Suggested that there are insects that lay eggs in other insects.
13	**circa 1675**	Otto Marsilius (Holland, 1619 - 1678)	Correspondence to Swammerdam	Informed Swammerdam how parasitoid eggs are laid in host insects.
14	**1678**	Jan Swammerdam and Otto Marsilius (Holland)	Book of Nature	Provided the first European description of the complete life cycle of a parasitoid based on comprehensive observation; published posthumously in 1738.
15	**1679**	Maria Sybilla Merian (Germany-Holland, 1647 - 1717)	Der Raupen wunderbare Verwandelung	Observed and illustrated emergence of parasitoid larvae from caterpillars.
16	**1685**	Martin Lister (Britain, 1639 - 1712)	De Insectis	Supposed that larvae observed by Goedaert emerging from caterpillars had developed from eggs laid by another insect.
17	**1685-1691**	Maria Sybilla Merian (Germany-Holland, 1647 - 1717)	Der Raupen wunderbare Verwandelung, Final Version	In the preface, provided a correct interpretation of insect parasitism based on observation of egg laying by parasitoids during 1685-1691; published posthumously in 1717.
18	**1686**	Marcello Malpighi (Italy, 1628 - 1694)	Opera omnia	Observed emergence of parasitoids but did not understand the phenomenon of insect parasitism.
19	**1687**	Antoni van Leeuwenhoek (Holland, 1632 - 1723)	Letter 59	Observed larvae and adult parasitoids, supposed they developed from eggs laid in or on the host by a parasitoid; expressed similar opinions in later letters but did not witness egg laying for a long time.

20	1690-1705	John Ray (Britain, 1627 - 1705)	Historia Insectorum	Correctly interpreted the phenomenon of insect parasitism but did not observe egg laying; his interpretation was published posthumously in 1710.
21	1692	Diacinto Cestoni (Italy, 1637 - 1718)	Letter to Vallisnieri	Described an attack of a whitefly by a parasitoid.
22	1696	Antonio Vallisnieri (Italy, 1661 - 1730)	Dialoghi, sopra la curiosa origine di molti insetti	Published a correct interpretation of insect parasitism but had not yet observed oviposition by a parasitoid.
23	1700	Antoni van Leeuwenhoek (Holland, 1632 - 1723)	Letter 134	Described in great detail the observation of oviposition and whole development of a parasitoid based on experimentation, providing a picture of a parasitoid in attack position.
24	1702	D. Nomoto (Japan, 1665 - 1714)	Methods for Sericulture	Mentioned a tachinid parasitoid of silkworm but lacked knowledge of its biology.
25	1704	Pu Songling (China, 1640 - 1715)	Works of Mr. Liao Zai - Notes after Disaster	Observed emergence of a hymenopteran parasitoid from a caterpillar; did not see oviposition, likely the first Chinese documentation of a hymenopteran parasitoid.
26	1717	Maria Sybilla Merian (Germany-Holland, 1647 - 1717)	Der Raupen wunderbare Verwandelung, Final Version	In the preface, described the full cycle of insect parasitism based on observations made between 1685-1691; this version was posthumously published in 1717.

Table: Timeline of Key Research and Discoveries in Biological Control

Year	Discovery/Event	Description
1888	**Introduction of *Rodolia cardinalis***	The vedalia beetle was introduced to California from Australia to control the cottony cushion scale, marking a significant early success in biological control.
1919	**Establishment of the first research organizations**	Governmental research agencies dedicated to biological control began to be established, fostering scientific exploration and implementation of biological control methods.
1926	**Discovery of *Bacillus thuringiensis* (Bt)**	The bacterium *Bacillus thuringiensis*, which produces toxins harmful to insects, was identified, later becoming a cornerstone of microbial pest control.

1940s	**Development of the sterile insect technique (SIT)**	The sterile insect technique was developed as a method of biological control, using radiation to sterilize pests, which are then released to reduce pest populations through failed reproductions.
1960s	**Integration into IPM**	Biological control strategies began to be integrated into integrated pest management (IPM) programs, emphasizing the use of multiple methods for pest control.
1971	**Establishment of International Organization for Biological Control (IOBC)**	The IOBC was founded to promote the study and implementation of biological control, enhancing international collaboration and standardization.
1980s	**Advances in genetic engineering**	Genetic engineering began to be applied to biological control, enhancing the specificity and effectiveness of biological control agents.
1990s	**Expansion of fungal biopesticides**	Research into entomopathogenic fungi expanded, leading to the development and commercialization of fungal strains as biopesticides.
2000s	**Use of RNA interference (RNAi)**	RNA interference technology was explored for its potential to silence specific genes in pests, offering a targeted approach to biological control.
2010s	**CRISPR and genome editing**	The advent of CRISPR and other genome-editing technologies opened new possibilities for modifying biological control agents for enhanced performance and safety.
2020s	**Artificial intelligence in biological control**	The integration of AI to model pest-predator interactions and optimize the deployment of biological control strategies represents the latest advancement in the field.

2. Principles of Biological Control

Biological control stands as a foundational strategy in sustainable pest management, using living organisms to regulate the populations of pest species. This approach not only helps maintain ecological balance but also reduces the reliance on chemical pesticides, which can have harmful effects on the environment. The core principles of biological control are rooted in ecology and evolutionary biology, focusing on the interactions between organisms. Among the main strategies in biological control, importation or classical biological control is pivotal. Here, we explore the concept and methodology of importation in detail.

a. Importation or Classical Biological Control

Concept and Historical Context

Importation, also known as classical biological control, involves introducing natural enemies from a pest's native range to a new environment where the pest has become invasive. This method is predicated on the observation that pests often become problematic in new geographic locations because they escape the natural predators that keep their populations in check in their native environments.

The modern era of biological control began with a landmark event in 1888 when the *Rodolia cardinalis* (vedalia beetle) was introduced to California from Australia to manage the cottony cushion scale. This successful application marked the beginning of what would become a globally recognized practice in pest management and set the stage for further developments in the field.

Process of Classical Biological Control

1. Pest Identification and Research

The initial phase of classical biological control involves identifying the invasive pest and conducting comprehensive research on its biology, behavior, and ecological role within its native habitat. This step is fundamental as it informs all subsequent actions in the biological control process. By understanding the pest's population dynamics, feeding habits, and reproductive strategies, researchers can predict how the pest might respond to various control methods and identify potential vulnerabilities.

The process begins with a thorough documentation of the pest's life cycle and behavior patterns under various environmental conditions. Specialists often collaborate with local experts and utilize historical data to assess the long-term patterns of pest outbreaks and their impacts on ecosystems. For example, understanding the breeding season of a pest can help in timing the introduction of natural enemies to maximize impact.

Laboratory and field studies are essential during this phase to simulate and observe the pest's behavior and its interaction with potential control agents. These studies help refine the selection of control methods by determining the most effective approaches based on the pest's natural history. For instance, if a pest is found to have a particular vulnerability during its larval stage, control efforts may focus on targeting this stage for maximum effectiveness.

2. Foreign Exploration

Foreign exploration involves conducting detailed surveys in the pest's native range to identify and collect potential natural enemies. This phase is crucial and requires an in-depth understanding of the local ecology and the specific interactions between the pest and its natural predators or pathogens.

Researchers undertake extensive field trips to the pest's native environment, where they observe the ecological interactions in situ. The goal is to discover and catalog natural enemies that are effectively controlling the pest in its original habitat. This could include predators, parasitoids, and pathogens that have evolved alongside the pest, thereby possessing innate capabilities to suppress its populations.

The selection of potential agents is guided by their demonstrated effectiveness in the field and their specificity to the target pest. This specificity is vital to ensure that the introduced agents do not become a threat to other native species. Field researchers often work in tandem with local biologists and ecologists to leverage regional expertise and resources, ensuring a comprehensive survey and collection process.

3. Evaluation and Selection of Natural Enemies

Once potential natural enemies are identified, they undergo a rigorous evaluation process to determine their suitability for release into the pest's new environment. This evaluation is critical to ensure that these agents are specific to the target pest and do not pose a risk to non-target species or the broader ecosystem.

The selection process involves several stages of testing in quarantine facilities, where the behavior, lifecycle, and host specificity of the agents are studied under controlled conditions. Researchers assess the potential for these agents to adapt to the target environment and their efficacy in controlling the pest population without causing unintended ecological consequences.

Bioassays are commonly used to test the effectiveness of the agents against the pest, while genetic and molecular techniques are employed to ascertain their specificity. The aim is to select agents that are highly specialized to the target pest, minimizing risks to other species. Environmental impact assessments are also performed to evaluate the potential ecological effects of introducing these agents.

4. Quarantine and Testing

Selected natural enemies undergo stringent quarantine procedures and further testing before they can be approved for field release. This phase is designed to mitigate the risk of inadvertently introducing other harmful organisms along with the control agents. Quarantine measures are meticulously followed to ensure the biological safety of the introduction.

During quarantine, the behavior and reproductive patterns of the agents are monitored in a secure environment. This allows scientists to observe the interactions between the agents and the target pest without external influences. Tests are conducted to verify that the agents do not carry diseases or parasites that could be transferred to native species.

The agents are also tested for their adaptability to the new environment's climatic and ecological conditions. This is crucial for ensuring their survival and effectiveness once released. Regulatory approvals are sought during this phase, with detailed documentation of the testing and evaluation results provided to governmental or international bodies overseeing biological control implementations.

5. Mass Production and Release

Successful completion of the quarantine and testing phase leads to the mass production of the chosen natural enemies. This step is vital to ensure that sufficient numbers of the agents are available to achieve control over the pest population. Production facilities are often specialized to the needs of the specific agents, providing the necessary conditions for breeding, such as controlled temperatures, humidity, and dietary requirements.

Release strategies are carefully planned based on the biology of the agent and the characteristics of the pest problem. Techniques such as inundative or inoculative releases are chosen depending on the desired impact and the lifecycle of the control agent. Inundative releases involve flooding the area with large numbers of agents to quickly reduce the pest population, while inoculative releases aim to establish a sustainable population of agents that can provide long-term control.

The release is often done in phases and monitored closely to assess initial success and to make necessary adjustments. Information from the release phase is crucial for refining future control efforts and for understanding the dynamics of agent-pest interactions in the field.

6. Monitoring and Evaluation

After the release, ongoing monitoring and evaluation are critical to assess the effectiveness of the biological control effort. This involves regular surveys of both the pest and the natural enemy populations to gather data on their interaction dynamics and population levels.

Monitoring techniques might include field sampling, remote sensing, and population modeling to obtain accurate measures of control efficacy. Feedback from these observations helps determine if the natural enemies are establishing themselves, spreading, and effectively controlling the pest population.

Adjustments to the management strategy might be necessary based on the results of this monitoring. For instance, additional releases may be needed, or habitat management practices might be adjusted to better support the natural enemies. Evaluation also includes assessing the impact on non-target species and the overall biodiversity of the area to ensure that the biological control intervention does not disrupt the ecological balance.

Through these detailed and carefully managed phases, classical biological control provides a sustainable approach to managing invasive pest species while minimizing the ecological footprint of pest control measures.

Ethical and Ecological Considerations

Implementing importation requires careful consideration of ethical and ecological impacts. The potential for non-target effects, where the introduced species negatively impacts native species or ecosystems, is a significant concern. Comprehensive risk assessment and ongoing monitoring are essential to mitigate these risks. The history of biological control includes both notable successes and instances where introduced agents have had unintended ecological impacts.

Examples of Successful Classical Biological Control

Classical biological control has demonstrated significant successes over the decades, with numerous well-documented cases showing how the introduction of natural enemies can lead to the sustainable management of invasive species. This section details three landmark examples: the control of the prickly pear cacti by *Cactoblastis cactorum* in Australia, the management of invasive European rabbits with the Myxoma virus, and the use of *Aphidius colemani*, a parasitic wasp, in controlling aphid populations across various agricultural settings.

1. Control of Prickly Pear Cacti by *Cactoblastis cactorum*

In the early 20th century, Australia faced an ecological and agricultural crisis due to the explosive spread of prickly pear cacti, introduced into the country for farming purposes. By the 1920s, these cacti had infested millions of hectares, rendering land unusable for agriculture and livestock grazing. The situation prompted one of the most notable early successes in classical biological control with the introduction of the moth *Cactoblastis cactorum* from South America, where it naturally preyed on prickly pear cacti.

Biological Mechanism and Implementation

Cactoblastis cactorum larvae burrow into cactus pads, feeding on the plant tissues and eventually causing a systemic collapse of the cactus structure. The biological control program began with careful research and selection of this moth, considering its host specificity to ensure that it would not adversely affect native or economically important plant species. Following quarantine and preliminary testing, the moths were released in affected areas.

Outcomes and Impact

The introduction of *Cactoblastis cactorum* led to a dramatic reduction in prickly pear populations across Australia. Within a decade, vast areas of land were reclaimed for use in agriculture and cattle farming, significantly boosting the Australian economy and altering the ecological landscape. The success of this program is often celebrated for its direct benefits and its role in promoting the wider acceptance and implementation of biological control strategies globally.

2. Management of Invasive European Rabbits with Myxoma Virus

The European rabbit, introduced into Australia in the late 19th century for hunting, quickly became an invasive species due to its high reproductive rate and lack of natural predators, leading to severe ecological and agricultural damage. In the 1950s, the Myxoma virus, a naturally occurring pathogen found in South American rabbits, was introduced as a biological control agent.

Biological Mechanism and Implementation

The Myxoma virus causes myxomatosis, a disease that is typically lethal in European rabbits. The virus was initially spread through direct injections into wild rabbit populations, and later through the distribution of insects such as mosquitoes and fleas that could transmit the disease. The implementation was carefully monitored to assess the spread and impact of the virus on rabbit populations and to ensure that it did not jump to non-target species.

Outcomes and Impact

The release of the Myxoma virus resulted in significant reductions in the European rabbit population, with mortality rates initially exceeding 90% in infected populations. This drastic decrease allowed native vegetation to regenerate and reduced the rabbits' impact on crops. However, over time, resistance to the virus began to develop in the rabbit population, leading to a gradual rebound in their numbers. This outcome has led to ongoing discussions about the ethics and long-term viability of using pathogens as biological control agents.

3. Use of *Aphidius colemani* to Control Aphid Populations

Aphids are major pests in various agricultural settings, capable of causing significant damage to crops by sucking sap from plants and transmitting plant diseases. *Aphidius colemani*, a parasitic wasp, has been employed successfully to manage aphid populations through biological control.

Biological Mechanism and Implementation

Aphidius colemani targets specific aphid species, injecting its eggs into the aphid's body. The hatching larvae consume the aphid from the inside, eventually killing it. This natural predator has been integrated into IPM programs worldwide, particularly in greenhouse environments where chemical controls can be undesirable.

Outcomes and Impact

The use of *Aphidius colemani* in agricultural systems has demonstrated significant reductions in aphid populations, leading to healthier crops and reduced chemical pesticide use. This strategy not only helps maintain the ecological balance but also supports sustainable agriculture by reducing the environmental footprint of pest management practices. The success of *Aphidius colemani* as a biological control agent is also a testament to the importance of selecting specific, effective natural enemies that do not adversely affect other species or the environment.

b. Augmentation

Augmentation, a principal strategy within biological control, involves the strategic addition of natural enemies to an environment to boost the existing populations' ability to control pests. This method is distinguished by its focus on supplementing rather than introducing new species to an ecosystem. There are two main forms of augmentation: inoculative and inundative releases, each tailored to different pest management goals and ecological contexts.

Inoculative Release: Long-term Pest Management

Inoculative release involves the periodic release of small numbers of natural enemies with the expectation that they will establish, persist, and exert control over pest populations over time. This strategy is particularly suited to agricultural or natural settings where long-term pest management is desired and where it is feasible for the natural enemies to survive and reproduce.

Concept and Application

The inoculative approach is often used when dealing with pests that have a steady presence and require ongoing control. It is particularly effective in perennial systems like orchards, vineyards, and forests. The process begins with identifying a compatible natural enemy that can adapt to the target environment and sustain its population without continuous human intervention.

The success of an inoculative release largely depends on the careful selection of the release time and the biological characteristics of the control agent. Factors such as the life cycle of the pest, the reproductive rate of the natural enemy, and the climatic conditions play crucial roles in determining the effectiveness of this strategy.

Benefits

The benefits of inoculative releases include:

- **Sustainability:** Once established, the natural enemies can provide ongoing control, reducing the need for repeated interventions.
- **Cost-effectiveness:** After initial establishment, the maintenance cost of pest control is generally low.
- **Environmental Impact:** Reduces the need for chemical pesticides, enhancing biodiversity and ecosystem health.

Challenges

However, inoculative release also faces several challenges:

- **Establishment Failure:** Natural enemies may fail to establish due to unsuitable environmental conditions, inadequate initial population sizes, or lack of sufficient food resources.
- **Slow Impact:** The effects on pest populations are not immediate, and it may take several seasons before significant impact is visible.

Inundative Release: Immediate Pest Reduction

In contrast to inoculative release, inundative release involves the mass release of natural enemies to achieve immediate pest control, akin to using a biopesticide. This method is typically used in situations where rapid pest suppression is needed or where long-term establishment of natural enemies is not viable.

Concept and Application

Inundative release is akin to treating natural enemies as a living pesticide. Large quantities of predators, parasitoids, or pathogens are released to overwhelm pest populations quickly. This approach is common in greenhouse agriculture or in annual crop systems where quick results are essential to protect a season's yield.

The production of natural enemies for inundative release requires specialized facilities where agents are bred in large numbers under controlled conditions. The timing and method of release are critical, with considerations for the pest's life cycle and environmental conditions to maximize the effectiveness of the control agents.

Benefits

The primary benefits of inundative release include:

- **Immediate Impact:** Provides rapid reduction in pest populations, which is crucial in acute infestation scenarios.
- **Flexibility:** Can be used as part of an integrated pest management strategy, combined with other biological, chemical, or cultural controls.
- **Safety:** Generally safer for humans and non-target organisms compared to chemical pesticides.

Challenges

Challenges associated with inundative releases include:

- **High Cost:** Mass production and release of biological agents can be expensive.
- **Reapplication Necessity:** Unlike inoculative releases, inundative releases do not typically result in long-term control, often requiring repeated applications.
- **Risk of Non-target Effects:** Although generally safe, there is a risk that released agents might affect non-target species, especially if not properly tailored to the pest.

Advantages of Biological Control for Farmers and Consumers

Biological control offers numerous benefits to both farmers and consumers by providing a safer, more sustainable alternative to chemical pesticides. Here's a detailed look at why farmers choose biological control and its advantages:

1. Reduced Exposure to Toxic Chemicals
    - Farmers and spray personnel face significantly lower health risks since biological control methods reduce the need for toxic chemical pesticides. This reduced exposure enhances safety during pest management activities.
2. **No Residues on Marketed Products**
    - Products treated with biological control agents typically have no chemical residues, making them safer and more appealing to health-conscious consumers. This lack of residues also meets the stringent standards required for organic certification.

3. **Protection of Plant Health**
   - Biological control agents are selected for their specific target pests and do not harm the plants themselves. This means there are no phytotoxic effects, such as damaged young plants or premature abortion of flowers and fruits. In many cases, this can lead to increased yields due to healthier plant growth.
4. **Ease and Pleasantness of Application**
   - Applying natural enemies in settings like greenhouses is less time-consuming and more pleasant compared to handling chemicals, especially in warm and humid conditions. This ease of use makes biological control a preferable option for many growers.
5. **Minimal Monitoring Required**
   - Once natural enemies are released and established shortly after planting, they require only occasional monitoring to ensure their effectiveness. This contrasts with chemical methods that need continuous oversight to manage application rates and timing effectively.
6. **Overcoming Pesticide Resistance**
   - Some pests have developed resistance to conventional pesticides, making chemical control difficult or ineffective. Biological control offers a viable alternative, targeting these resistant pests without the drawbacks of chemical dependency.
7. **No Safety Interval Required**
   - Unlike chemical pesticides, which often require a waiting period between application and harvest to ensure safety, biological control allows for immediate harvesting. This is particularly advantageous for crops with fluctuating market prices, as it provides flexibility in timing the harvest to maximize profit.
8. **Long-term Sustainability**
   - Once an effective natural enemy is established, it remains a reliable ally against pests. Biological control is not a one-time solution but a permanent one, reducing the need for repeated applications and thereby lowering long-term management costs.
9. **Public Appreciation and Market Benefits**
   - The general public increasingly prefers products that are environmentally friendly and sustainably produced. Crops grown under biological control are often perceived as healthier and safer, which can lead to quicker sales, better prices, or both.

c. Conservation in Biological Control

Conservation, a key strategy in biological control, focuses on enhancing the effectiveness of natural enemies already present in the environment. This is achieved by modifying agricultural or natural practices to create a more favorable habitat for these beneficial organisms. By supporting natural predators and parasitoids, conservation biological control aims to increase their impact on pest populations, thereby reducing the need for chemical interventions.

Part 1: Principles and Approaches in Conservation Biological Control

Principles of Conservation Biological Control

The foundational principle of conservation biological control is that natural ecosystems possess inherent mechanisms for pest control through predator-prey relationships and parasitism. Agricultural systems, however, often lack the diversity and structural complexity of natural ecosystems, which can diminish the effectiveness of natural enemies. Conservation biological control seeks to reintroduce or enhance these ecological functions by altering environmental conditions or farming practices to support the life cycles and predatory efficacy of natural enemies.

Key Approaches in Conservation

1. **Habitat Manipulation:** This involves modifying the agricultural environment to make it more conducive for natural enemies to thrive. Techniques include planting cover crops, maintaining hedgerows, and creating beetle banks. These practices provide shelter, alternative prey, and overwintering sites for predators and parasitoids, thereby enhancing their survival and reproduction rates.

2. **Reduced Pesticide Use:** Chemical pesticides can have detrimental effects on non-target organisms, including natural enemies. Reducing pesticide use, either through decreased application, targeted application, or substitution with biopesticides, can significantly benefit natural predator populations. Integrated Pest Management (IPM) programs often incorporate these strategies to optimize pest control while minimizing environmental impact.

3. **Providing Supplementary Resources:** Supplying natural enemies with additional resources can help boost their populations. This might include providing nectar sources through flowering plants, which sustain adult parasitoids and predators, or supplying artificial honeydew and pollen.

4. **Fostering Landscape Diversity:** A diverse landscape with a mix of crops, non-crop areas, and natural habitats supports a wider range of natural enemies and reduces the likelihood of pest outbreaks. Landscape diversity can be enhanced by incorporating buffer zones, maintaining natural vegetation patches, and practicing crop rotation.

Benefits of Conservation Biological Control

- **Sustainability:** Enhances biodiversity and promotes the stability of agricultural ecosystems.
- **Cost-effectiveness:** Reduces the need for expensive chemical inputs over time.
- **Long-term pest control:** Supports sustainable pest suppression across multiple growing seasons.
- **Reduced environmental impact:** Minimizes chemical runoff and preserves soil health.

Part 2: Implementing Conservation Strategies

Steps for Implementing Conservation Biological Control

Implementing conservation strategies requires a comprehensive approach that integrates knowledge of ecology, agronomy, and pest management. Here's a structured way to approach this:

1. **Assessment of Current Practices:** Begin by evaluating the existing agricultural or land management practices to identify how they might be impacting natural enemies. This assessment should consider pesticide usage, crop rotation schedules, and existing habitat structures.
2. **Identification of Key Natural Enemies:** Identify which natural enemies are present in the system and their specific needs. This might involve field surveys or consulting with entomologists and local agricultural extension services.
3. **Designing a Conservation Plan:** Based on the assessment, develop a plan that addresses the needs of the natural enemies. This might include creating specific habitats, adjusting pesticide applications, or introducing new agricultural practices like intercropping or cover cropping.
4. **Implementation of Habitat Enhancements:** Execute the habitat modifications or enhancements as planned. This could involve planting nectar-rich border plants, establishing permanent grassy banks, or restoring wetlands adjacent to agricultural fields.
5. **Monitoring and Adjustment:** Once the conservation measures are in place, continuous monitoring is essential to evaluate their effectiveness

and make necessary adjustments. Monitoring should focus on the populations of both pests and natural enemies, and adjustments might include further modifications to habitat or changes in crop management practices.

Challenges in Conservation Biological Control

While conservation biological control offers numerous benefits, it also faces several challenges:

- **Time and Knowledge:** Implementing conservation strategies requires a significant investment in time and a deep understanding of ecological dynamics, which may be a barrier for some farmers and land managers.
- **Initial Costs:** Some habitat enhancements, such as establishing hedgerows or cover crops, require upfront investments which may not provide immediate economic returns.
- **Complex Interactions:** The ecological interactions in a given area can be complex, making it difficult to predict the outcomes of conservation efforts. Unintended consequences, such as the support of non-target pests, can occur.

Future Prospects

The future of conservation biological control looks promising as awareness of the environmental impacts of traditional pest control methods grows. Advances in ecological research, as well as greater collaboration between farmers, scientists, and policymakers, are likely to drive the development of more effective conservation strategies. As the agricultural community continues to embrace sustainable practices, conservation biological control will play a crucial role in shaping resilient agricultural ecosystems that can withstand the challenges of pests and environmental changes.

3. Scope of Biological Control

Biological control, as a discipline within agricultural and environmental sciences, has grown to become a cornerstone of integrated pest management (IPM) and sustainable agricultural practices. Over the decades, the scope of biological control has expanded significantly, encompassing a wide range of organisms and strategies that target not only agricultural pests but also invasive species that threaten natural ecosystems.

Broadening Horizons: Diverse Agents and Targets

Initially focused on the use of predatory insects and parasitoids, the scope of biological control has broadened to include mites, nematodes, and an array of microbial pathogens such as bacteria, fungi, and viruses. Each group of organisms offers unique mechanisms for controlling pest populations, providing multiple layers of intervention within IPM systems.

1. **Insects and Mites**: Predatory insects such as lady beetles, lacewings, and hoverflies, and mites have been used to control aphids, mites, and other destructive pests. These agents are often specific to their target pests, minimizing the impact on non-target species.
2. **Nematodes**: Beneficial nematodes attack soil-dwelling insect larvae, offering a way to control subterranean pests without the need for chemical soil treatments. They are particularly effective against grubs, weevils, and other root-feeding insects.
3. **Microbial Pathogens**: Bacteria like *Bacillus thuringiensis* (Bt) produce toxins that are lethal to certain insects but harmless to others, including humans. Fungi such as *Beauveria bassiana* and *Metarhizium anisopliae* infect and kill a variety of insects, while viruses specific to insects can cause disease outbreaks in pest populations, leading to their collapse.

The integration of these diverse agents into a coherent pest management strategy allows for more effective and sustainable pest control, reducing the reliance on chemical pesticides which can lead to resistance and environmental degradation.

Integration into Integrated Pest Management (IPM)

Biological control is increasingly recognized as an integral component of IPM strategies, which aim to use a combination of biological, physical, chemical, and cultural methods to manage pest populations in an economically and ecologically rational manner. The integration of biological control enhances IPM by:

- **Reducing Chemical Use**: By relying on natural predators and pathogens, IPM systems can minimize the use of chemical pesticides, thereby reducing environmental toxicity and the risk of developing pest resistance.
- **Enhancing Long-term Sustainability**: Biological control agents often establish themselves in the ecosystem, providing ongoing control without continual human intervention, which is vital for sustainable pest management.
- **Improving Crop Yields and Quality**: Effective biological control improves plant health by reducing pest damage, which in turn can increase crop yields and improve the quality of agricultural products.

Biotechnological Advances

Advancements in biotechnology have dramatically influenced the scope and efficacy of biological control. Genetic engineering and other biotechnological tools are being used to enhance the specificity, efficiency, and safety of biological control agents:

- **Genetic Modification**: Genetic engineering can enhance the ability of biocontrol agents to resist diseases, tolerate harsh environmental conditions, or increase their effectiveness against specific pests.
- **RNA Interference (RNAi)**: This technique can silence specific genes in target pests, making them more vulnerable to biological control agents or less capable of causing harm.
- **Precision Breeding**: Advances in genomics and breeding technologies enable the development of hyper-efficient biocontrol agents tailored for specific environmental conditions or pest challenges.

Public Perception and Regulatory Advances

The public perception of biological control has improved as awareness of the environmental impacts of pesticides has increased. Consumers are more supportive of sustainable practices in agriculture, which has led to greater acceptance and demand for products grown with biological control methods.

Regulatory frameworks have also evolved to support the use of biocontrol agents. Governments and international bodies are establishing more rigorous testing and approval processes to ensure that biological control practices are safe, effective, and beneficial. These regulations help maintain biodiversity and protect ecosystems while supporting agricultural productivity.

Table: Key Biological Control Agents and Their Targets

S No.	Biological Agent	Target Pests	Example of Use
1	*Trichogramma spp.*	Various Lepidoptera larvae	Widely used in agriculture for moth control.
2	*Bacillus thuringiensis*	Caterpillars, beetles, flies	Employed as a biopesticide in various crops.
3	*Coccinellidae* (ladybeetles)	Aphids, scale insects	Used for aphid control in gardens and fields.
4	*Encarsia formosa*	Whiteflies	Utilized in greenhouse settings for whitefly control.
5	*Phytoseiulus persimilis*	Spider mites	Employed in greenhouses to manage spider mite infestations.
6	*Aphidoletes aphidimyza*	Aphids	Applied in both greenhouses and outdoor crops.

7	*Beauveria bassiana*	Thrips, whiteflies, aphids	Used as a fungal biopesticide in various settings.
8	*Metarhizium anisopliae*	Beetles, ticks, termites	Used in forestry and agriculture to control soil pests.
9	*Steinernema carpocapsae*	Caterpillars, cutworms, root weevils	Employed in turf management and vegetable crops.
10	*Heterorhabditis bacteriophora*	Grubs, weevils	Used in orchards and gardens to control soil-dwelling pests.
11	*Goniozus legneri*	Moth larvae	Used in nut orchards to control navel orangeworms.
12	*Nasonia vitripennis*	Fly pupae	Utilized in livestock settings to control house flies.
13	*Muscidifurax raptor*	Fly pupae	Employed in animal husbandry to reduce fly populations.
14	*Anagyrus pseudococci*	Vine mealybug	Used in vineyards to manage mealybug infestations.
15	*Diatomaceous earth*	Various crawling insects	Used as a mechanical pesticide in grains and storage facilities.

4. Future Perspectives in Biological Control

The field of biological control is at a transformative juncture, with rapid advancements in technology and shifts in agricultural practices shaping its evolution. The convergence of molecular biology, biotechnology, and ecological awareness is paving the way for innovative approaches that promise to enhance the efficacy, specificity, and sustainability of biological control methods. As global agriculture increasingly leans towards sustainability, the role of biological control in managing pest populations and preserving ecological integrity is becoming more critical than ever.

Advances in Molecular Biology and Biotechnology

1. **RNA Interference (RNAi)**: RNAi technology, which allows for the silencing of specific genes in target organisms, is revolutionizing pest control strategies. By targeting genes essential for pest survival, reproduction, or virulence, RNAi provides a highly specific way to weaken or eliminate pests without harming non-target species. This method is not only more environmentally friendly but also reduces the chances of pest resistance compared to traditional chemical pesticides.
2. **CRISPR and Genome Editing**: The advent of CRISPR-Cas9 and other genome-editing tools offers unprecedented precision in modifying the genetic makeup of organisms. In biological control, CRISPR can be used to enhance the traits of natural enemies, making them more robust, effective, or specific in their actions. Additionally, genome editing could

be employed to develop sterile insect techniques or to engineer insects that are incapable of transmitting diseases.

3. **Synthetic Biology**: This field combines biology and engineering to construct new biological parts, devices, and systems or to redesign existing, natural biological systems for useful purposes. In biological control, synthetic biology could lead to the creation of novel biocontrol agents with enhanced capabilities or to the development of microbial consortia that act synergistically against pests.

Regulatory Frameworks and Policy Support

As biotechnological innovations in biological control advance, regulatory frameworks are also evolving to address the new challenges and opportunities these technologies present. Regulators are increasingly focusing on ensuring that biocontrol agents and their applications are safe for the environment, non-target organisms, and human health. Enhanced regulatory protocols are being developed to evaluate the ecological impacts, potential risks, and benefits of genetically modified organisms (GMOs) used in biological control.

Moreover, policy support for biological control is growing as governments recognize its value in sustainable agriculture and biodiversity conservation. This is evident in the increasing funding for research in biological control, the integration of biological control strategies into national agricultural policies, and the support for international collaborations in research and development.

Public Acceptance and Market Dynamics

Public perception of biological control has significantly improved with greater awareness of environmental issues and the adverse effects of chemical pesticides. Consumers are more supportive of eco-friendly agricultural products and practices, which drives demand for crops produced with biological control methods. This shift is encouraging farmers to adopt biological control solutions, and it is influencing market dynamics in the agricultural sector.

The future market for biological control looks robust, with projections indicating growth driven by increased demand for organic products, regulatory pressures on chemical pesticides, and technological innovations that make biological control more effective and cost-competitive.

Environmental and Ecological Implications

Biological control is aligned with the principles of ecological conservation and sustainability. It plays a pivotal role in preserving biodiversity by maintaining natural predator-prey relationships and reducing the reliance on chemical interventions that can disrupt ecological balances. As climate change and environmental degradation become more pressing, biological control offers a resilient pest management strategy that can adapt to changing environmental conditions and help stabilize agro-ecosystems.

B. Important Groups of Parasitoids, Predators, and Pathogens in Biological Control

Biological control utilizes a diverse array of organisms to manage pest populations in agricultural and natural ecosystems.

1. Parasitoids in Biological Control

Parasitoids play a crucial role in the regulation of pest populations through biological control. They are unique among biological control agents in that they have a very specific life cycle involving the use of a host organism, in which they eventually kill the host, making them effective agents for pest management.

Hymenoptera

The order Hymenoptera includes some of the most important and diverse groups of parasitoids used in biological control, such as Braconidae, Ichneumonidae, and Chalcididae. Each family contains species that target specific pest insects, making them invaluable for integrated pest management (IPM) strategies.

1. **Braconidae**: This family comprises a large number of species known for their role in controlling a variety of agricultural pests. Braconid wasps are typically small to medium in size and usually attack caterpillars and beetle larvae. For instance, *Cotesia congregata* is a braconid wasp that parasitizes the tobacco hornworm, a common pest on tobacco and tomato plants. The female wasp injects her eggs into the hornworm, and the emerging larvae feed internally, eventually leading to the host's death.
2. **Ichneumonidae**: One of the largest families of Hymenoptera, ichneumonid wasps are highly specialized parasitoids, many of which are used to control forest pests and other insect populations that affect trees and shrubs. An example is *Ophion*, which lays its eggs in the bodies of large caterpillars. After hatching, the larvae consume the caterpillar from the inside out.
3. **Chalcididae**: Chalcid wasps are smaller than many other parasitoids but are significant in controlling populations of scale insects, whiteflies, and other pests that affect fruits and ornamental plants. *Encarsia formosa*, a well-known chalcid wasp, is effectively used against whiteflies in greenhouse environments.

4. **Trichogrammatidae**: *Trichogramma spp.* are minute wasps that parasitize a broad range of lepidopteran eggs, including those of many harmful moths and butterflies. These wasps are extensively used worldwide to control agricultural pests. The female wasp injects her eggs into the eggs of the pest, where the larvae develop, preventing the pest eggs from hatching.

Diptera: Tachinid Flies in Biological Control

Tachinid flies are another vital group within the biological control arsenal. These flies belong to the order Diptera and are known for their parasitic relationship with host insects, particularly caterpillars and beetle larvae.

1. **Life Cycle and Mechanism**: Tachinid flies lay their eggs on the bodies of host insects. Once the eggs hatch, the larvae penetrate the host's body and begin to consume it from the inside, eventually leading to the host's death. This method of parasitism makes tachinid flies effective for controlling large populations of pests.
2. **Examples of Tachinid Flies**: One of the best-known tachinid flies is *Compsilura concinnata*, introduced to North America to control gypsy moths and brown-tail moths. Another example is *Archytas marmoratus*, which targets a variety of caterpillar species that are pests on crops such as corn, cotton, and tomatoes.

Application and Effectiveness

The application of parasitoids in biological control programs involves careful planning and consideration of the ecological balance in the target area. Effective use of parasitoids requires knowledge of their life cycles, host preferences, and environmental requirements.

1. **Release Strategies**: Inoculative and augmentative releases are common strategies. Inoculative release involves introducing a small population of parasitoids into the environment, expecting them to establish and control the pest population over time. Augmentative release may be periodic or inundative, involving larger numbers of parasitoids released to quickly reduce a pest population.
2. **Challenges and Considerations**: While parasitoids are powerful tools for biological control, their effectiveness can be influenced by several factors, including climatic conditions, availability of hosts, and the presence of other natural enemies. Additionally, the risk of non-target effects must be carefully managed to prevent ecological imbalances.

2. Predators in Biological Control

Predators play an indispensable role in natural and managed ecosystems by controlling pest populations and maintaining ecological balance. Their ability to consume multiple prey items throughout their lifetimes makes them critical allies in sustainable agricultural practices, particularly in integrated pest management (IPM) systems.

1. Coccinellidae (Ladybeetles)

Ladybeetles, or ladybugs, are one of the most recognizable and beneficial predatory insects, celebrated for their appetite for aphids and other pest insects. Their contribution to biological control is substantial, owing to their generalist feeding habits and adaptability to various environments.

Biology and Ecology

- Ladybeetles belong to the family Coccinellidae, which includes thousands of species worldwide.
- They are known for their rounded bodies and colorful patterns, often red or yellow with black spots.
- Ladybeetles undergo complete metamorphosis, with four life stages: egg, larva, pupa, and adult.

Feeding Habits

- Both larval and adult stages of ladybeetles are predatory, with aphids being their primary food source.
- A single ladybeetle can consume up to 5,000 aphids in its lifetime, making it an extremely effective agent for controlling aphid populations in gardens, greenhouses, and crops.

Species Examples and Uses

- *Coccinella septempunctata*, the seven-spot ladybird, is widely used across Europe and North America due to its efficiency in controlling aphid populations in a variety of crops.
- *Harmonia axyridis*, although effective, has also become an invasive species in many regions, illustrating the need for careful management and selection of species for biological control programs.

Table: Common Ladybeetle Species Used in Biological Control

Species	Target Pest	Common Usage
Coccinella septempunctata	Aphids	Crop fields, gardens
Harmonia axyridis	Aphids, scale insects	Gardens, orchards, greenhouses
Adalia bipunctata	Aphids, mites	Urban green spaces, greenhouses

2. Araneae (Spiders)

Spiders are generalist predators that play a vital role in controlling a variety of insect pests in and around crop fields. They are particularly valued for their capacity to manage pest populations without specific intervention, serving as a natural pest control agent in many agricultural and garden settings.

Biology and Ecology

- Spiders belong to the order Araneae and are found in nearly every habitat on Earth, including agricultural fields.
- They are not insects but arachnids, characterized by their eight legs and chelicerae (mouthparts with fangs that inject venom).

Feeding Habits

- Spiders are opportunistic predators, feeding on a wide range of insects, including many pest species such as flies, moths, mosquitoes, and beetles.
- They utilize various hunting strategies, from spinning webs to trap prey to actively pursuing or ambushing their targets.

Role in Agriculture

- By reducing pest populations, spiders contribute to the suppression of pest outbreaks and help reduce the need for chemical pesticides.
- Their presence in a field can be an indicator of environmental health and a well-functioning ecosystem.

3. Chrysopidae (Green Lacewings)

Green lacewings are another cornerstone of biological control strategies, particularly valued for their larvae's aggressive predation of aphids, thrips, and other pest insects.

Biology and Ecology

- Green lacewings are insects in the family Chrysopidae, known for their delicate, green wings and golden eyes.
- They undergo complete metamorphosis, with the larvae stages being the most predatory.

Feeding Habits

- Lacewing larvae are known as "aphid lions" for their voracious appetite for aphids and other soft-bodied insects.
- Adults typically feed on nectar, pollen, and honeydew, although some species are predatory as well.

Use in IPM

- Lacewings are used extensively in both agricultural and garden settings for their effectiveness in controlling a variety of pests.
- They are particularly effective in greenhouse environments where their natural tendencies can be maximized in controlled conditions.

Table: Lacewing Species Commonly Used in Biological Control

Species	Target Pest	Common Usage
Chrysoperla carnea	Aphids, mites, thrips	Greenhouses, vegetable crops
Chrysoperla rufilabris	Aphids, caterpillars	Orchards, vineyards

Table: Predatory Species Commonly Used in Biological Control

S No.	Predator Species	Target Pest	Common Usage
1	*Orius insidiosus*	Thrips, aphids, mites, small caterpillars	Greenhouses, strawberries, cotton
2	*Podisus maculiventris*	Various caterpillars, beetle larvae	Vegetable crops, field crops like soybeans
3	*Nabis spp.*	Aphids, caterpillars, leafhoppers	Alfalfa, cotton, soybeans
4	*Hippodamia convergens*	Aphids, scale insects	Gardens, orchards, greenhouses
5	*Geocoris spp.*	Aphids, spider mites, caterpillars, eggs	Cotton, corn, and other field crops

3. Pathogens in Biological Control

Pathogens play a crucial role in biological control by causing diseases that naturally manage pest populations. This method involves utilizing bacteria, fungi, viruses, and nematodes that are specifically harmful to pests but are generally safe for humans, beneficial insects, and the environment. Each type of pathogen has unique characteristics and methods of infecting and killing host pests.

1. Bacteria

Bacillus thuringiensis (Bt)

Bacillus thuringiensis, commonly known as Bt, is a soil-dwelling bacterium that produces proteins (endotoxins) which form crystals harmful to many insect species. These proteins specifically target the gut lining of insects, causing them to starve and eventually die after ingestion.

- **Mechanism**: When Bt is ingested by susceptible insect larvae, the alkaline pH of their gut activates the toxin. The toxin then binds to specific receptors on the gut cells, creating pores that cause cell lysis and eventual death of the larvae.
- **Usage**: Bt strains are formulated into various biopesticide products suitable for spraying on foliage. They are used against a wide range of lepidopteran (moths and butterflies), coleopteran (beetles), and dipteran (flies and mosquitoes) pests. Bt is particularly favored in organic farming and is effective in controlling pests like corn borer, cotton bollworm, and gypsy moth caterpillars.
- **Safety and Environmental Impact**: Bt is non-toxic to humans, animals, and beneficial insects, making it an environmentally friendly alternative to chemical pesticides.

Bacillus sphaericus

- **Mechanism**: *Bacillus sphaericus* produces toxins that, like Bt, are harmful to specific insects. This bacterium is particularly effective against mosquito larvae. The toxins disrupt the digestive systems of the larvae, leading to starvation and death.
- **Usage**: It is commonly used in mosquito control programs, especially in areas prone to malaria and other mosquito-borne diseases. The bacterium is formulated into granules or liquids that can be applied to water bodies where mosquito larvae are present.
- **Safety and Environmental Impact**: *Bacillus sphaericus* is highly specific to mosquito larvae and does not affect beneficial insects, humans, or other wildlife, making it an environmentally safe option.

Bacillus popilliae (often referred to as the causative agent of "milky spore disease")

- **Mechanism**: The bacterium produces spores that infect and kill larvae of the Japanese beetle. Infected larvae often show a characteristic milky white appearance due to bacterial growth.

- **Usage**: *Bacillus popilliae* is specifically used for controlling Japanese beetle populations in lawns, gardens, and golf courses where these pests are a significant problem.
- **Safety and Environmental Impact**: This bacterium is safe for humans, pets, and beneficial insects, focusing its pathogenic effects solely on Japanese beetle larvae.

Paenibacillus popilliae (formerly known as *Bacillus popilliae*)

- **Mechanism**: Similar to *Bacillus popilliae*, *Paenibacillus popilliae* targets Japanese beetles by producing spores that infect the grubs, leading to a fatal disease known as milky spore disease.
- **Usage**: Used predominantly in turf management and in areas where Japanese beetles damage ornamental plants.
- **Safety and Environmental Impact**: The targeted action against Japanese beetles makes it safe for use around humans and animals, without impacting non-target species.

Bacillus thuringiensis israelensis **(Bti)**

- **Mechanism**: *Bti* is a strain of *Bacillus thuringiensis* that produces toxins effective specifically against larvae of blackflies and mosquitoes. These toxins disrupt the gut lining of the larvae, causing death.
- **Usage**: *Bti* is widely utilized in vector control programs to reduce populations of mosquitoes and blackflies, thereby helping to control diseases like malaria and river blindness.
- **Safety and Environmental Impact**: *Bti* is non-toxic to non-target organisms, including beneficial insects, fish, birds, and mammals, making it ideal for application in sensitive aquatic environments.

Bacillus thuringiensis kurstaki **(Btk)**

- **Mechanism**: A specific strain of *Bacillus thuringiensis*, *Btk* produces crystal proteins (Cry toxins) that are toxic when ingested by larvae of various moth and butterfly species.
- **Usage**: *Btk* is extensively used in forestry and agriculture to control caterpillar pests that damage crops and trees, such as the gypsy moth and tent caterpillars.
- **Safety and Environmental Impact**: *Btk* is highly specific to its target pests and is considered safe for humans, larger animals, and the environment. It is particularly favored in areas where chemical pesticides pose a risk to ecological health or where organic farming practices are in place.

2. Fungi

Beauveria bassiana and *Metarhizium anisopliae*

Fungi such as *Beauveria bassiana* and *Metarhizium anisopliae* are widespread soil fungi known for their ability to infect and kill various insect pests.

- **Mechanism**: These fungi grow naturally in soils worldwide and attack their hosts by penetrating the insect cuticle using mechanical pressure and enzymatic degradation. Once inside, they proliferate inside the host's body, producing toxins and draining the nutrients, resulting in the host's death.
- **Usage**: These fungal pathogens are applied as spore suspensions or powders in areas where insects are likely to come into contact with them. They are effective against a range of pests, including whiteflies, thrips, termites, and beetles.
- **Benefits**: The fungi offer prolonged pest control as they can persist in the environment and infect subsequent generations of pests. They are especially valuable in controlling pests that live in the soil or have become resistant to chemical insecticides.

Streptomyces avermitilis

- **Mechanism**: Produces avermectins, a group of compounds with potent insecticidal and nematocidal properties. Avermectins disrupt neural and muscular functions in targeted pests.
- **Usage**: Used primarily to control nematodes and insect pests in various crops. It's also the source of the commercial product Avermectin, which is widely used to control parasitic infections in livestock.
- **Safety and Environmental Impact**: Generally safe for humans and non-target animals at used concentrations, but care must be taken to ensure it does not affect beneficial insects and soil biota.

Pseudomonas fluorescens

- **Mechanism**: Produces a variety of secondary metabolites that inhibit fungal growth and suppress diseases. It also induces systemic resistance in plants, enhancing their ability to fend off diseases.
- **Usage**: Widely used to suppress soil-borne pathogens and promote plant growth by enhancing nutrient uptake. It is effective against diseases such as take-all in wheat and damping-off in seedlings caused by Pythium spp.

- **Safety and Environmental Impact**: Non-toxic to humans and animals, and beneficial for promoting soil health by enhancing nutrient cycling.

Serratia entomophila

- **Mechanism**: Produces toxins that specifically target the gut of the New Zealand grass grub, causing the insects to stop feeding, which eventually leads to their death.
- **Usage**: Applied in pastures to control grass grub populations, which can cause significant damage to grasslands and cereal crops by feeding on their roots.
- **Safety and Environmental Impact**: Highly specific to its target pest and poses no known risks to humans, non-target organisms, or the environment.

Burkholderia spp.

- **Mechanism**: Some species produce antibiotics and other bioactive metabolites that suppress pathogenic fungi and bacteria. They also promote plant growth and induce systemic resistance against a variety of pathogens.
- **Usage**: Used in bioformulations to protect crops from bacterial and fungal pathogens. Effective against rice bacterial wilt and banana Xanthomonas wilt.
- **Safety and Environmental Impact**: Generally safe, though some strains of Burkholderia are pathogenic to humans, and careful strain selection and management are essential to ensure safety.

Lysobacter enzymogenes

- **Mechanism**: Known for producing a range of enzymes and antibiotics that degrade the cell walls of fungi and other bacteria. This bacterium provides excellent control against fungal pathogens.
- **Usage**: Utilized to combat fungal diseases in crops, such as damping-off and root rot, particularly in greenhouse environments where humidity can exacerbate fungal growth.
- **Safety and Environmental Impact**: Considered safe for use around humans and animals and is beneficial for integrated pest management programs focusing on reducing fungicide use.

3. Viruses

Nucleopolyhedro Viruses (NPVs)

Insect-specific viruses like nucleopolyhedro Viruses are pathogens that specifically infect and kill certain insects, particularly caterpillars of moths and butterflies.

- **Mechanism**: NPVs are ingested by the host insect during feeding. The virus particles infect cells lining the midgut, multiply inside the host, and systematically infect other tissues. Infected larvae eventually die and disintegrate, releasing new virus particles into the environment.
- **Usage**: NPVs are applied as biopesticides, particularly in forestry and agricultural settings, to control major pests such as the gypsy moth and forest tent caterpillar.
- **Advantages**: These viruses are highly specific to their hosts, which minimizes impact on non-target species. Their use in large-scale forestry operations demonstrates their effectiveness in managing pest outbreaks over wide areas.

Granuloviruses (GVs)

- **Mechanism**: Similar to NPVs, Granuloviruses (GVs) are ingested by the host caterpillar, where the virus particles infect the midgut cells, replicate extensively, and eventually cause systemic infection that leads to the host's death.
- **Usage**: GVs are particularly effective against smaller caterpillar populations and are often used in horticulture and smaller crop fields. For example, the Codling Moth Granulovirus (*Cydia pomonella* GV) is extensively used in apple orchards to control the codling moth, a major pest of apples.
- **Advantages**: GVs have a very narrow host range, which significantly reduces any risk to non-target organisms, including beneficial insects, wildlife, and humans.

Cydia pomonella Granulovirus (CpGV)

- **Mechanism**: CpGV specifically targets the larvae of the codling moth, a common pest in apple and pear orchards. The virus causes a disease that slows larval development, leading to death before the larvae can cause significant damage to the fruit.
- **Usage**: This virus is widely used in integrated pest management programs in orchards to reduce reliance on chemical pesticides and manage resistance issues.

- **Advantages**: CpGV is highly effective in controlling codling moth populations and has an excellent safety profile, with no adverse effects on humans, animals, or the environment.

Helicoverpa armigera Nucleopolyhedro Viruses (HaNPV)

- **Mechanism**: HaNPV specifically infects the cotton bollworm (*Helicoverpa armigera*), a significant pest on various crops worldwide. The virus disrupts the normal cellular processes of the larvae, causing death.
- **Usage**: HaNPV is used in various agricultural settings, particularly in cotton, corn, and tomato crops, where *Helicoverpa armigera* poses a significant threat.
- **Advantages**: The use of HaNPV helps in significantly reducing the bollworm populations without the environmental and health risks associated with chemical insecticides.

Spodoptera exigua Nucleopolyhedro Viruses (SeNPV)

- **Mechanism**: SeNPV infects the beet armyworm, *Spodoptera exigua*, a common pest in vegetable and ornamental crops. The virus multiplies inside the larvae, causing disease and death.
- **Usage**: It is applied as a foliar spray in affected fields, particularly in intensive vegetable production areas.
- **Advantages**: SeNPV is host-specific and does not affect non-target insects or other wildlife, making it a safe and environmentally friendly option for pest control.

Autographa californica M Nucleopolyhedro Viruses (AcMNPV)

- **Mechanism**: AcMNPV infects a wide range of larvae from different moth species, making it one of the more versatile NPVs. It causes a fatal disease that disrupts normal feeding and development.
- **Usage**: Although it has a broader host range, AcMNPV is primarily used against species like the alfalfa looper and other related pests in a variety of crops.
- **Advantages**: Its ability to control multiple pest species reduces the need for multiple agents, simplifying management practices and reducing overall pest control costs.

4. Nematodes

Nematodes in Biological Control: Steinernema and Heterorhabditis

Entomopathogenic nematodes, particularly those from the genera *Steinernema* and *Heterorhabditis*, play a critical role in the biological control of various pest species, primarily soil-dwelling insects. These microscopic worms are highly effective due to their unique infection mechanism and symbiosis with lethal bacteria, making them an invaluable tool in sustainable agriculture.

Mechanism of Action

Entomopathogenic nematodes invade their host insects through natural body openings like the mouth, anus, and respiratory spiracles, or by directly penetrating the body wall. Once inside the host, nematodes release symbiotic bacteria from their digestive tracts into the insect's blood. These bacteria, primarily from the genera *Xenorhabdus* and *Photorhabdus*, multiply rapidly, producing toxins that lead to sepsis and rapid death of the host, usually within 48 hours.

The effectiveness of this mechanism lies in its specificity and the lethal efficiency of the bacterial symbionts, which are adapted to act quickly to outcompete the host's immune response and prevent recovery.

Usage in Pest Management

Entomopathogenic nematodes are particularly valued for their ability to target and control a variety of pest insects that are otherwise difficult to manage due to their subterranean habitats:

1. **Root-feeding insects**: Nematodes are used to manage pests like white grubs, root weevils, and other soil-dwelling larvae that cause significant damage to crops such as potatoes, strawberries, and ornamental plants.
2. **Stem and wood borers**: Certain species of nematodes are effective against the larvae of borers that affect trees and shrubs.
3. **Foliage pests**: Although primarily soil-dwellers, nematodes can also be used against surface-dwelling pests such as caterpillars and cutworms when applied during moist conditions that favor nematode survival on plant surfaces.

Application techniques include soil drenches, seed treatments, and foliar applications, often using irrigation systems or standard pesticide sprayers adapted for nematode dispersion.

Benefits of Using Nematodes

The use of entomopathogenic nematodes in pest control offers several significant benefits:

- **Safety**: Nematodes are safe for humans, animals, and beneficial insects, making them an excellent choice for integrated pest management programs in agricultural, urban, and forested environments.

- **Environmental Impact**: Unlike chemical pesticides, nematodes do not leave harmful residues in the environment, nor do they affect non-target species, ensuring minimal ecological disruption.
- **Versatility**: They can be used in a variety of settings, from greenhouses and nurseries to fields and orchards.
- **Resistance Management**: Nematodes provide an alternative mode of action to help manage resistance development in pest populations that may be becoming resistant to chemical insecticides.

Steinernema and Heterorhabditis

Entomopathogenic nematodes are microscopic worms that infect and kill insects, primarily those living in the soil.

- **Mechanism**: These nematodes enter the host body through natural body openings (mouth, anus, respiratory spiracles) or sometimes directly through the body wall. Once inside, they release symbiotic bacteria from their gut, which quickly multiply, poisoning the blood and killing the host within 48 hours.
- **Usage**: They are applied to the soil to control root-feeding insects like grubs and weevils, and are also effective against caterpillars and cutworms.
- **Benefits**: Nematodes are safe for humans and animals, highly specific to targeted pests, and can be used with other biological control agents.

Steinernema carpocapsae

- **Target Pests**: Known for its effectiveness against a wide range of soil-dwelling and foliar-feeding pests, including armyworms, cutworms, and root weevils.
- **Mechanism**: Known as an "ambush" predator, this nematode species waits for its prey near the soil surface and attacks rapidly when a host comes into contact.
- **Usage**: Widely used in vegetable and ornamental plant production, particularly effective in managed soil systems and containerized plant settings.

Steinernema feltiae

- **Target Pests**: Effective against thrips, some root weevils, and fungus gnats.
- **Mechanism**: This nematode actively searches out its prey rather than waiting for it to pass by, making it effective against pests in thicker soil or those that burrow deeper.

- **Usage**: Commonly used in greenhouse environments where humidity levels aid in its survival and movement, particularly for ornamental crops and dense vegetable plots.

Heterorhabditis bacteriophora

- **Target Pests**: Primarily targets soil-dwelling insects like the Japanese beetle, citrus root weevils, and various white grubs.
- **Mechanism**: This nematode moves actively through soil and locates hosts by sensing temperature gradients and carbon dioxide emissions from host insects.
- **Usage**: Applied in agricultural and lawn care programs to control grub populations that can cause significant damage to roots and turf.

Steinernema glaseri

- **Target Pests**: Known for its efficacy against larger soil pests such as mole crickets, Japanese beetles, and other large grubs.
- **Mechanism**: Similar to other Steinernema species, it uses a cruise strategy to locate hosts but is particularly adept at infecting larger insects.
- **Usage**: Useful in sports fields, pasture lands, and other large turf areas where mole crickets and grubs are common pests.

Steinernema riobrave

- **Target Pests**: Targets a variety of soil-dwelling pests including corn rootworms, citrus root weevils, and other beetles.
- **Mechanism**: This nematode is adapted to hotter climates and can survive in lower moisture conditions compared to other nematodes.
- **Usage**: Often used in arid and semi-arid regions where its drought tolerance makes it a viable option for biological pest control.

C. History of insect pathology

Table: Broader spectrum of developments in insect pathology

S No	Year	Development/Discovery	Examples/Details
1	1668	Francesco Redi proposes spontaneous generation	Challenged the notion of insects spontaneously generating from decay.
2	1743	John Needham's experiments	Suggested support for spontaneous generation with boiled broth.
3	1765	Lazzaro Spallanzani's experiments	Disproved spontaneous generation through sealed flask experiments.
4	1835	Agostino Bassi discovers Beauveria bassiana	First recognized insect-pathogenic fungus.
5	1859	Louis Pasteur's experiments	Further discredited spontaneous generation with swan-necked flasks.
6	1879	Heinrich de Bary identifies Entomophthora muscae	Recognized as the first identified entomopathogenic fungus.
7	1884	Metchnikoff's studies on insect immunity	Pioneered the study of insect immune responses.
8	1911	Hans Zinsser identifies Rickettsia	Rickettsial pathogens discovered, some affecting insects.
9	1915	Carlos Chagas discovers Trypanosoma cruzi	Agent of Chagas disease, transmitted by insects.
10	1920	Friedrich Henle isolates the first baculovirus	Baculovirus isolated from the larvae of Galleria mellonella.
11	1926	Paul Müller discovers the insecticidal properties of DDT	DDT revolutionizes insect control but later banned due to environmental concerns.
12	1934	A. W. Bacot and C. J. Martin identify arboviruses in mosquitoes	Recognized mosquitoes as vectors for arboviruses.
13	1946	Nobel Prize in Physiology or Medicine for insect physiology	Awarded to André Michel Lwoff, Jacques Monod, and François Jacob.
14	1951	First recorded use of Bacillus thuringiensis (Bt) as an insecticide	Widely used bioinsecticide for controlling lepidopteran pests.
15	1956	Development of the sterile insect technique (SIT)	Method for controlling insect populations through sterilization.
16	1963	Discovery of Wolbachia bacteria in insects	Significant for its reproductive manipulation effects on insects.
17	1978	First molecular studies of insect viruses	Initiation of genetic research on insect pathogens.
18	1982	Discovery of insect antiviral immune responses	Understanding insect immune defenses against viruses.
19	1991	Development of transgenic plants expressing insecticidal proteins	Offers alternative pest control methods.

20	**1999**	Complete genome sequence of Drosophila melanogaster	Milestone in understanding insect genetics and immunity.
21	**2003**	Discovery of RNA interference (RNAi) in insects	Revolutionized insect pest control through gene silencing.
22	**2013**	First evidence of horizontal gene transfer between insects and viruses	Implications for insect immune response and virus evolution.
23	**2018**	CRISPR gene editing adapted for use in insects	Opens up possibilities for precise manipulation of insect genomes.
24	**2020**	Emergence of novel insect pests and diseases due to climate change	Increased spread and impact of insect-borne diseases.
25	**2022**	Development of insect-resistant genetically modified crops	Addresses agricultural pest challenges while reducing pesticide use.

D. Infection of insects by bacteria, fungi, viruses, protozoa, rickettsiae, spiroplasma and nematodes

S No	Pathogen Type	*Species Name*	Mode of Action	Target Insects	Plants Affected
1	**Bacteria**	***Bacillus thuringiensis***	Produces toxins lethal to insects by disrupting gut	Lepidopteran larvae (e.g., caterpillars)	Various crops (e.g., corn, cotton)
2		***Bacillus sphaericus***	Produces mosquitocidal toxins targeting larval midgut	Mosquito larvae (e.g., *Culex*, *Anopheles* species)	Standing water habitats (e.g., ponds)
3		***Pseudomonas entomophila***	Produces toxins that disrupt gut epithelium	Fruit flies (e.g., *Drosophila melanogaster*)	Fruit orchards, vineyards
4		***Xenorhabdus nematophila***	Produces toxins that kill insect hosts	Nematodes (e.g., *Steinernema*, *Heterorhabditis* species)	Various crops, soil environments
5		***Serratia entomophila***	Produces toxins that disrupt insect immune system	Beetles (e.g., Colorado potato beetle)	Potato, tomato crops
6		***Yersinia pestis***	Causes septicemia and death in insect hosts	Fleas (e.g., *Xenopsylla cheopis*)	Various mammals, humans
7		***Photorhabdus luminescens***	Produces toxins that disrupt insect hemocoel	Insect larvae (e.g., moth larvae)	Soil environments
8		***Chromobacterium piscinae***	Produces toxins lethal to insect larvae	Mosquito larvae (e.g., *Aedes aegypti*)	Water bodies, wetlands
9	**Fungi**	***Beauveria bassiana***	Adheres to insect cuticle, penetrates integument	Beetles, caterpillars, aphids	Various crops, forests
10		***Metarhizium anisopliae***	Adheres to insect cuticle, produces toxins	Grasshoppers, termites, whiteflies	Various crops, forests
11		***Cordyceps militaris***	Invades insect body, manipulates behavior	Ants, caterpillars, beetles	Forests, grasslands
12		***Isaria fumosorosea***	Penetrates insect cuticle, infects internal tissues	Whiteflies, thrips, aphids	Greenhouses, orchards

13		***Entomophaga maimaiga***	Infects and kills insect larvae by consuming internal tissues	Gypsy moth larvae	Forests, urban areas
14		***Hirsutella thompsonii***	Infects and kills insect hosts by penetrating cuticle	Mites, beetles, aphids	Various crops, forests
15		***Lecanicillium lecanii***	Penetrates insect cuticle, infects and kills host	Whiteflies, aphids, scale insects	Greenhouses, orchards
16		***Nomuraea rileyi***	Produces toxins that disrupt insect gut epithelium	Caterpillars, beetles, grasshoppers	Various crops, forests
17	**Viruses**	***Baculovirus***	Infects midgut cells, causes systemic infection	Lepidopteran larvae (e.g., cabbage moth)	Various crops (e.g., cabbage, corn)
18		***Densovirus***	Disrupts host developmental processes	Mosquitoes, black flies, sandflies	Various plants, aquatic environments
19		***Iridovirus***	Infects various tissues, causes systemic infection	Aquatic insects (e.g., dragonflies, damselflies)	Wetlands, ponds, streams
20		***Dicistrovirus***	Infects midgut cells, disrupts cellular function	Honey bees, bumblebees, solitary bees	Various crops, flowering plants
21		***Reovirus***	Infects midgut cells, causes tissue damage	Leafhoppers, planthoppers, aphids	Various crops, grasslands
22		***Nudivirus***	Infects various tissues, alters host physiology	Beetles, caterpillars, moths	Various plants, forests
23		***Flavivirus***	Disrupts insect immune responses, causes systemic infection	Mosquitoes, ticks, sandflies	Various plants, wetlands
24		***Cypovirus***	Infects midgut cells, interferes with nutrient absorption	Grasshoppers, locusts, crickets	Grasslands, agricultural fields
25	**Protozoa**	***Nosema apis***	Invades midgut epithelium, disrupts cellular function	Honey bees, bumblebees, moths	Various crops, flowering plants
26		***Crithidia bombi***	Infects gut cells, disrupts nutrient uptake	Bumblebees, solitary bees, moths	Various flowering plants, forests

27		***Trypanosoma cruzi***	Invades various tissues, disrupts host physiology	Triatomine bugs, kissing bugs, assassin bugs	Various mammals, birds, reptiles
28		***Leishmania major***	Infects various tissues, causes tissue damage	Sandflies, mosquitoes, black flies	Various mammals, birds, reptiles
29		***Plasmodium falciparum***	Invades salivary glands, disrupts feeding behavior	Mosquitoes (e.g., *Anopheles* species)	Various mammals, birds, reptiles
30		***Toxoplasma gondii***	Invades brain tissue, alters behavior	Beetles, flies, grasshoppers	Various mammals, birds, reptiles
31		***Giardia lamblia***	Invades midgut epithelium, disrupts nutrient absorption	Flies, mosquitoes, beetles	Various mammals, birds, reptiles
32		***Cryptosporidium parvum***	Invades midgut epithelium, disrupts cellular function	Mosquitoes, black flies, sandflies	Various mammals, birds, reptiles
33	**Nematodes**	***Steinernema carpocapsae***	Infective juveniles penetrate cuticle, release bacteria	Beetle larvae, caterpillars, fly larvae	Various crops, forests
34		***Heterorhabditis bacteriophora***	Infective juveniles penetrate cuticle, release bacteria	White grubs, caterpillars, root maggots	Various crops, soil environments
35		***Steinernema feltiae***	Infective juveniles penetrate cuticle, release bacteria	Fungus gnats, thrips, caterpillars	Greenhouses, orchards
36		***Heterorhabditis indica***	Infective juveniles penetrate cuticle, release bacteria	Lepidopteran larvae (e.g., cabbage moth)	Various crops, forests
37		***Steinernema glaseri***	Infective juveniles penetrate cuticle, release bacteria	Beetle larvae, caterpillars, fly larvae	Various crops, forests
38		***Heterorhabditis megidis***	Infective juveniles penetrate cuticle, release bacteria	White grubs, caterpillars, root maggots	Various crops, soil environments
39		***Steinernema feltiae***	Infective juveniles penetrate cuticle, release bacteria	Fungus gnats, thrips, caterpillars	Greenhouses, orchards
40		***Heterorhabditis bacteriophora***	Infective juveniles penetrate cuticle, release bacteria	White grubs, caterpillars, root maggots	Various crops, soil environments

1. **Bacteria:** *Bacillus thuringiensis* (Bt) and *Bacillus sphaericus* are two notable examples of bacteria with entomopathogenic properties. Bt produces toxins lethal to insects by disrupting their gut epithelium, primarily affecting lepidopteran larvae such as caterpillars. On the other hand, *B. sphaericus* targets mosquito larvae, including species like Culex and Anopheles, by producing mosquitocidal toxins that disrupt larval midgut function. These bacteria are widely used in agriculture and vector control programs, respectively, due to their specificity to target pests and minimal impact on non-target organisms.
2. **Fungi:** *Beauveria bassiana* and *Metarhizium anisopliae* are prominent entomopathogenic fungi that infect a variety of insect hosts. *B. bassiana* adheres to the insect cuticle, penetrating the integument and causing mycosis, affecting beetles, caterpillars, and aphids. Similarly, *M. anisopliae* produces toxins upon contact with the insect cuticle, targeting grasshoppers, termites, and whiteflies. These fungi are valuable in biological control strategies for managing insect pests in agricultural and forestry settings, providing an alternative to chemical pesticides.
3. **Viruses:** Baculoviruses, such as those affecting lepidopteran larvae, and densoviruses, impacting mosquitoes and black flies, are examples of insect-infecting viruses. Baculoviruses infect midgut cells, causing systemic infection in target insects like cabbage moths, while densoviruses disrupt host developmental processes, affecting a variety of insects including mosquitoes and sandflies. These viruses have been explored as biocontrol agents for managing insect populations in agricultural systems and natural habitats, offering environmentally friendly alternatives to traditional pest management approaches.
4. **Protozoa:** Protozoan parasites like Nosema apis and *Crithidia bombi* infect insect hosts, disrupting cellular functions and impacting their physiology. *Nosema apis* invade the midgut epithelium of honey bees, affecting their nutrient absorption and overall health. Similarly, *Crithidia bombi* infects the gut cells of bumblebees, leading to reduced nutrient uptake and compromised immune responses. These protozoa can influence insect population dynamics and community interactions, particularly in pollinator species crucial for ecosystem functioning and crop pollination.
5. **Nematodes:** Entomopathogenic nematodes such as *Steinernema carpocapsae* and *Heterorhabditis bacteriophora* are parasitic worms that infect insect hosts, releasing symbiotic bacteria to kill the host. These nematodes penetrate the insect cuticle, targeting larvae of various

pests like beetles, caterpillars, and fly larvae. They are used in biological control programs to manage insect pests in agriculture, horticulture, and forestry, providing a sustainable and environmentally friendly alternative to chemical insecticides.

E. Plant Protection–Entomology

Plant Protection in Entomology is a crucial discipline within agricultural sciences focused on the management and control of insect pests to safeguard crop health and productivity. This field is integral to ensuring food security and promoting sustainable agricultural practices worldwide. It encompasses a variety of research areas, technologies, and strategic applications aimed at minimizing pest-induced damages while reducing environmental impacts and conserving beneficial organisms.

Understanding Insect Pests

Biology and Behavior

- Detailed study of insect pests includes understanding their life cycles, which vary widely among species and can influence the timing and method of interventions. For instance, knowing that certain pests, like the corn borer, have specific breeding times allows farmers to time their pest control measures more effectively.
- Researchers also focus on the feeding habits and reproductive behaviors of pests. For example, aphids reproduce asexually and rapidly during warm months, necessitating quick and effective control measures to prevent outbreaks.

Ecology

- The ecological study of pest interactions with their environment includes their relationships with host plants and natural enemies. This holistic view helps in devising pest management strategies that are ecologically sound and sustainable.
- Understanding ecological factors also involves studying how environmental changes, such as climate change, impact pest dynamics and management.

Methods

Biological Control

- **Introduction of Natural Enemies**: This involves using predators, parasitoids, and pathogens to naturally suppress pest populations. For example, releasing lady beetles to control aphid populations on crops.

- **Microbial Control**: The application of microbial agents like *Bacillus thuringiensis* is an effective strategy against various larvae, which consume the bacteria and succumb to the toxins produced within their gut.

Chemical Control

- **Target-Specific Pesticides**: These pesticides are designed to target specific pests to reduce non-target effects. For example, neonicotinoids are used against sucking pests but are applied in ways that minimize exposure to beneficial insects.
- **Integrated Pest Management (IPM)**: IPM strategies integrate biological, chemical, and cultural practices to manage pests in an economically and ecologically rational manner. This includes crop rotation, the use of resistant varieties, and precise pesticide application.

Advantages and Significance

Sustainable Agriculture

- Plant protection methods help in maintaining sustainable agricultural systems by enhancing biodiversity and reducing reliance on chemical pesticides.
- These practices contribute to soil health, reduce pollution, and help in the conservation of natural resources.

Food Security

- By reducing crop damages and improving plant health, effective pest management directly contributes to higher agricultural productivity and stability in food supplies.

Economic Benefits

- Effective pest management not only saves costs associated with crop damage but also reduces expenditure on chemical pesticides. Sustainable practices enhance the marketability of agricultural products in markets increasingly sensitive to environmental and health concerns.

Technological Advances

Genetic Engineering

- Genetic modifications of crops to express pest-resistant traits offer a promising avenue to reduce the need for external chemical inputs. For example, Bt corn expresses a bacterial protein that kills specific pests, thereby reducing the need for insecticides.

Precision Agriculture

- Advanced technologies such as drones, satellite imaging, and IoT devices are revolutionizing pest monitoring and control, enabling more precise and efficient pest management practices.

Biotechnology

- Developments in microbial pesticides and genetically engineered natural enemies are enhancing the specificity and effectiveness of biological control agents.

Future Prospects

Climate Change Adaptation

- Ongoing research is crucial for developing pest management strategies that can adapt to changing climatic conditions, which influence pest populations and their geographic distributions.

Biological Innovations

- The exploration of new biological control agents and techniques, such as RNA interference and gene editing technologies, holds promise for innovative approaches to pest management.

Data-Driven Solutions

- Leveraging big data analytics and machine learning to predict pest outbreaks and optimize management strategies could significantly enhance the responsiveness and effectiveness of pest control measures.

Public Awareness and Policy Support

- Increased public awareness and robust policy frameworks are essential to promote the adoption of sustainable pest management practices and support research and innovation in plant protection.

Short Questions

1. What is the principle of classical biological control and how does it differ from augmentation?
2. List three important groups of parasitoids used in biological control.
3. What role do nematodes play in the biological control of insect pests?
4. Identify two pathogens that are commonly used in microbial control of pests.
5. What is the significance of *Bacillus thuringiensis* in plant protection entomology?

Long Questions

1. Discuss the history and development of biological control from its origins to modern practices. Include key milestones in the evolution of this field and how these developments have influenced current pest management strategies.
2. Examine the roles and mechanisms of three major types of biological control agents: parasitoids, predators, and pathogens. Provide examples of each and describe how they are typically integrated into agricultural pest management programs.
3. Describe the principles of Integrated Pest Management (IPM) and explain how conservation, augmentation, and importation strategies are implemented within an IPM framework. Include examples of how each strategy is used in practice and discuss the benefits and challenges associated with these approaches.

Unit II

Biology, adaptation, host seeking behaviour of predatory and parasitic groups of insects. Role of insect pathogenic nematodes, viruses, bacteria, fungi, protozoa, etc., their mode of action. Biological control of weeds using insects. Epizootiology, symptomatology and etiology of diseases caused by the above and the factors controlling these. Defense mechanisms in insects against pathogens.

A. Biology, Adaptation, and Host-Seeking Behavior of Predatory and Parasitic Insects

The study of the biological processes, adaptations, and host-seeking behaviors of predatory and parasitic insects is fundamental for enhancing the efficacy of pest management strategies. These aspects are central to leveraging natural predator and parasite behaviors for biological control.

i. Biology of Predatory and Parasitic Insects

The study of the biology of predatory and parasitic insects is a cornerstone in understanding how these organisms function as critical components of biological control programs. These insects play a pivotal role as natural enemies of pest species, with physiological and developmental traits finely tuned for the effective management of pest populations. By examining their complex life cycles, reproductive strategies, and other biological traits, researchers can optimize the use of these insects in integrated pest management (IPM) strategies.

Life Cycles of Predatory and Parasitic Insects

The life cycle of an insect is a fundamental aspect of its biology that influences its role in ecosystem dynamics, particularly in pest control. Predatory and parasitic insects typically undergo a holometabolous life cycle, which includes four distinct stages: egg, larva, pupa, and adult. Each stage presents unique characteristics and vulnerabilities that can be targeted by pest management professionals to enhance control strategies.

1. Egg Stage

The egg stage is the initial phase of the life cycle of predatory and parasitic insects and is crucial for establishing future generations. Because eggs are stationary and clustered, they are vulnerable to targeted pest management practices. One effective method is the use of ovicides, chemicals specifically designed to kill eggs. However, in integrated pest management (IPM) programs, more sustainable approaches such as habitat manipulation or disruption are preferred.

Habitat Manipulation: Modifying the environment to make it less hospitable for egg laying and survival can significantly decrease pest populations. This can involve altering moisture levels, removing plant debris that may provide shelter, or using cover crops that deter pest insects from laying eggs.

Biological Control: Introducing or enhancing natural enemies that specifically target eggs can also be an effective strategy. Certain species of beetles and flies are known to prey on the eggs of pest species, and promoting their presence can help keep pest populations in check.

These methods not only prevent the emergence of the next generation of pests but also reduce the need for more aggressive chemical interventions later in the pest's life cycle.

2. Larval Stage

The larval stage is when most of the feeding and growth occur, making it a critical time for pest management. Predatory insects like lady beetles and lacewings are especially effective during this stage, as they consume large quantities of soft-bodied pests such as aphids and scale insects. Parasitic insects, such as parasitoid wasps, target this stage by laying eggs inside or on the larvae, eventually killing the host as they develop.

Enhancing Natural Predators: Encouraging the presence of natural predators through habitat enhancement or supplementary feeding can increase the natural control of pests during this stage. This might involve planting nectar-producing flowers to attract adult predators or providing shelters for them to increase their survival and reproduction.

Biological Insecticides: Microbial products, such as those containing Bacillus thuringiensis (Bt), target specific larval stages of certain pests and can be applied as part of an IPM strategy. These products are generally safe for natural enemies and non-target species, making them an ideal choice for sustainable pest management.

By targeting the larval stage, pest management professionals can significantly reduce the damage pests cause to crops while promoting ecological balance.

3. Pupal Stage

The pupal stage, a period of immobility and vulnerability, offers a unique opportunity for pest control. During this stage, insects are encased within a pupal casing and are not feeding, which makes direct chemical treatments less effective but gives rise to other control options.

Physical and Mechanical Controls: These include techniques such as tillage, which can expose or destroy pupae in the soil, or using barriers to prevent emerging adults from reaching the soil surface to pupate.

Environmental Controls: Modifying environmental conditions to disrupt pupation can also be effective. This may involve altering humidity or temperature conditions to levels that are not conducive to successful pupation.

Addressing pests during the pupal stage can effectively decrease the next generation of pests, reducing the overall pest pressure and the need for interventions during the adult stage.

4. Adult Stage

Adult predatory and parasitic insects are integral to controlling pest populations, as they are often the most mobile and reproductive stage. Effective strategies targeting adults can have immediate impacts on pest populations and long-term benefits in reducing future generations.

Trapping and Monitoring: Traps baited with pheromones or other attractants can capture large numbers of adults, particularly in agricultural settings. Monitoring these traps can also provide valuable information on pest population dynamics, helping to inform the timing of other control measures.

Sterilization Techniques: Techniques such as the sterile insect technique (SIT) involve releasing sterilized males that compete with wild males for females but do not produce viable offspring, gradually reducing the population over time.

Regulatory and Behavioral Controls: Using insect growth regulators (IGRs) that interfere with the insect's ability to mature and reproduce can effectively reduce pest populations without harming non-target species.

ii. Reproductive Strategies of Predatory and Parasitic Insects

The reproductive strategies of predatory and parasitic insects are key to their success in controlling pest populations. These strategies are typically characterized by high fecundity and rapid population growth in response to the availability of prey or hosts.

1. High Fecundity

High fecundity, or the ability of organisms to produce a large number of offspring, is a fundamental trait in many parasitic and predatory insects that are utilized in biological control programs. This trait is particularly pronounced in parasitoid wasps, lady beetles, lacewings, and many other beneficial insects that play pivotal roles in managing pest populations in agricultural and natural ecosystems.

Fecundity in Biological Control Agents

Fecundity in biological control agents refers to their reproductive capacity, which is often a critical factor in their ability to effectively suppress pest populations. High fecundity rates can dramatically increase the efficiency and impact of biological control initiatives, ensuring that pest populations are kept under control through the continual replenishment of the beneficial insect populations.

1. **Biological Significance**: The high fecundity of these agents allows for rapid multiplication of beneficial populations in response to the availability of pests, which serve as food or hosts. This is particularly important in agroecosystems, where pest outbreaks can occur swiftly and can cause significant damage if not quickly contained.
2. **Ecological Dynamics**: In ecosystems, high fecundity helps maintain the balance between predator and prey populations. It ensures that beneficial insects can keep up with the reproductive rates of pest species, many of which also exhibit high fecundity rates. This natural balance is crucial for the stability of ecosystems and the prevention of any single species becoming overly dominant.

Mechanisms and Implications of High Fecundity

The ability of predatory and parasitic insects to lay a large number of eggs is influenced by several factors including genetics, environmental conditions, and the availability of resources. These mechanisms provides insights into how best to support these beneficial insects in various ecological settings.

1. **Genetic Factors**: Many parasitic and predatory insects have evolved over millions of years to maximize their reproductive success under varying environmental conditions. Genetic adaptations may promote faster development times, greater survivability of offspring, or more efficient use of resources, all of which can enhance fecundity.
2. **Environmental Influences**: Temperature, humidity, and the availability of food resources significantly affect reproductive rates. For instance,

warmer temperatures can often accelerate reproductive cycles and increase egg-laying rates, assuming other conditions such as food supply are optimal.

3. **Resource Availability**: The abundance of prey or host insects directly influences fecundity. Parasitoid wasps, for example, will lay more eggs if there are ample hosts available. Similarly, predatory insects like ladybeetles exhibit higher fecundity when aphid populations, their primary food source, are plentiful.

Practical Applications in Pest Management

The high fecundity trait of biological control agents has practical applications in agriculture, forestry, and urban pest management. It allows for sustainable pest control strategies that reduce reliance on chemical pesticides and promote ecological balance.

1. **Augmentative Releases**: In biological control programs, particularly augmentative releases where additional beneficial insects are released to boost natural populations, selecting species with high fecundity ensures that the introduced population can sustain itself and grow, even after the initial release period.
2. **Integrated Pest Management (IPM)**: High fecundity is a valuable trait in the context of IPM, where the goal is to integrate biological, chemical, and cultural control methods to manage pest populations effectively. Predatory and parasitic insects with high fecundity can rapidly respond to pest outbreaks, providing a dynamic and adaptable component to IPM strategies.
3. **Conservation Biological Control**: Enhancing the habitat to support high fecundity rates in natural populations of beneficial insects can be an effective long-term strategy. This might involve planting floral resources to provide nectar and pollen for adult insects, or maintaining mulch and other organic matter that fosters a conducive environment for their offspring.
2. **Host and Prey Selection**: The ability to choose specific hosts or prey is a refined trait among these insects. Parasitoids, for example, have evolved sophisticated mechanisms to detect host cues such as pheromones, vibrations, or excretions. This specificity helps in targeting pest species effectively without impacting non-target species, thus maintaining biodiversity and ecosystem health.

Advanced Sensory Mechanisms for Host and Prey Detection

Predatory and parasitic insects employ a variety of sensory mechanisms tailored to their ecological niches and target species. These mechanisms are adapted to detect subtle cues that guide these insects to their prey, optimizing their hunting efficiency and ensuring their survival.

1. **Chemical Cues**
   - **Pheromones**: Many insect pests release pheromones to attract mates, which can also attract parasitoids and predators. For example, parasitoid wasps are adept at detecting the specific pheromones released by caterpillars of certain moths, guiding them directly to their hosts. This precise chemical detection ensures that the biological control agents do not waste resources on non-target species.
   - **Volatile Organic Compounds (VOCs)**: Plants emit a variety of VOCs in response to pest attacks. These compounds can serve as indirect cues for predatory and parasitic insects, signaling the presence of prey. An example is the release of sesquiterpenes by maize plants when attacked by caterpillars, which specifically attracts parasitic wasps that prey on those caterpillars.
2. **Vibrational Cues**
   - Many parasitoids and predators can detect the vibrations caused by the feeding activity of their prey. This ability is particularly crucial for species like the emerald ash borer parasitoid, which can sense the feeding vibrations of larvae within the bark of trees. Such capabilities allow parasitoids to locate hidden prey, an advantage in environments where visual cues are obstructed.
3. **Visual and Physical Cues**
   - **Visual Detection**: Predators such as lady beetles utilize visual cues to spot aggregations of aphids on plant leaves. The ability to visually identify large groups of prey allows these predators to efficiently find and consume pests.
   - **Physical Detection**: Some predators detect the physical signs left by pests, such as silk trails or frass (insect feces), which can lead them directly to their targets. This type of cue is especially vital for predators that hunt in complex environments where chemical and vibrational cues might be less effective.

Ecological and Agricultural Implications of Host and Prey Selection

The precision with which predatory and parasitic insects select their targets has far-reaching implications for both natural ecosystems and agricultural settings.

1. **Biodiversity Conservation**
   - By targeting only specific pests, these biological control agents help maintain the ecological balance, avoiding the unintended consequences that broader-spectrum chemical pesticides might induce, such as the decline of non-target species including pollinators.
2. **Enhanced Agricultural Productivity**
   - In agricultural ecosystems, the specificity of these agents minimizes damage to crops and reduces the need for chemical interventions. This not only leads to healthier crop production but also supports sustainable agriculture practices by reducing chemical runoff and preserving soil health.
3. **Sustainable Pest Management**
   - The ability to selectively target pest species makes biological control a key component of Integrated Pest Management (IPM) strategies. This specificity is crucial for long-term sustainability, as it prevents the development of resistance that often results from the overuse of chemical pesticides.a

4. **Synchronized Emergence**
   - Many predatory and parasitic insects have synchronized their life cycles with those of their prey or hosts. This synchronization ensures that the most vulnerable stages of the pest are available when the parasitoids or predators are active. Techniques that manipulate environmental conditions can further enhance this synchronization, leading to more effective pest control.

1. **Maximal Impact**
   - When predators or parasitoids emerge simultaneously with a surge in the pest population, they can immediately begin feeding on or parasitizing these pests. This immediate interaction prevents the pests from reaching damaging population levels, thereby protecting crops and natural habitats from the adverse effects of pest outbreaks.
2. **Efficient Resource Use**
   - By ensuring that natural enemies are active during peak pest availability, these agents can find adequate food resources to sustain their populations. This not only enhances the survival and reproduction rates of the natural enemies but also increases their overall predatory or parasitic efficiency.

3. **Prevention of Pest Escapes**
   - Synchronizing the emergence of biological control agents with peak pest periods minimizes the risk of pests escaping control measures and establishing large populations. This is particularly important in agricultural settings where the timing of pest control can directly affect crop yields and quality.

Mechanisms Facilitating Synchronized Emergence

Achieving synchronized emergence involves a combination of biological insights and environmental manipulations to ensure that the life cycles of control agents are in harmony with pest dynamics.

1. **Environmental Cues**
   - Many biological control agents depend on temperature and photoperiod cues to regulate their development. Manipulating these environmental factors can accelerate or delay the development of these agents to better match the emergence of pests. For example, cooling storage areas can delay the development of parasitoid wasps until their host pests begin to appear in the fields.
2. **Genetic Selection**
   - Selective breeding programs can enhance traits in biological control agents that favor synchronization with pest life cycles. By selecting for agents that naturally emerge in sync with pest populations, researchers can create more effective biological control strains.
3. **Habitat Manipulation**
   - Modifying the habitat to support the lifecycle synchronization involves planting cover crops or maintaining buffer zones that provide resources for natural enemies at critical times. This strategic habitat management helps ensure that natural enemies are well-positioned to combat pests as soon as they appear.

Applications and Benefits

The concept of synchronized emergence is applied across various ecosystems, particularly in agriculture where timing pest control can significantly influence crop productivity.

1. **Agricultural Applications**
   - In crop systems, the synchronized release of egg parasitoids like Trichogramma spp. to target pest eggs during peak laying times can significantly reduce pest populations without the need for chemical interventions.

2. **Conservation Benefits**
   - Synchronizing the activity of natural enemies with pests helps maintain ecological balance and biodiversity by ensuring that control measures target only pest species and do not disrupt non-target species.
3. **Economic Advantages**
   - By reducing the reliance on chemical pesticides, synchronized emergence helps lower production costs, reduces environmental contamination, and meets consumer demands for sustainably produced agricultural products.

Challenges and Considerations

Despite its advantages, implementing synchronized emergence successfully presents several challenges:

1. **Timing Accuracy**
   - Accurately predicting the timing of pest and natural enemy life cycles requires detailed ecological knowledge and constant monitoring. Misalignments can lead to ineffective control and wasted resources.
2. **Pest Resistance**
   - There is a potential for pests to develop behavioral or physiological changes that desynchronize them from natural enemies, reducing the effectiveness of biological control strategies.
3. **Complexity in Implementation**
   - In open and uncontrolled environments, achieving precise synchronization can be challenging due to the myriad of uncontrollable factors that might influence the life cycles of both pests and natural enemies.

4. Dispersion and Migration in Biological Control Agents

Dispersion and migration are critical ecological behaviors of predatory and parasitic insects used in biological control. These processes play a significant role in the spatial dynamics of these agents and their effectiveness in controlling pest populations across various landscapes. These behaviors can greatly enhance the success of biological control programs by ensuring that control agents effectively colonize and manage pest populations over a wide area.

Understanding Dispersion and Migration

Dispersion refers to the spread of individuals across a landscape from their origin or release points, while migration involves the often-seasonal movement of populations from one region to another. Both behaviors are influenced by a range of ecological, environmental, and physiological factors.

1. **Ecological Significance**
   - Dispersion and migration enable biological control agents to reach pest populations that are dispersed across large agricultural fields or natural landscapes. This capacity ensures that these beneficial agents can locate and suppress pest outbreaks that occur away from their initial point of release or natural habitat.
2. **Adaptive Strategies**
   - These behaviors are adaptive responses to environmental variability, resource availability, and interspecific interactions. For instance, migration can occur in response to seasonal changes that affect the availability of food resources or suitable habitats.

Mechanisms Driving Dispersion and Migration

The mechanisms underlying dispersion and migration in biological control agents are complex and can include a combination of innate behaviors and environmental triggers.

1. **Innate Dispersal Mechanisms**
   - Many predatory and parasitic insects exhibit innate behaviors that promote dispersion. These can include random walking, directed movement towards favorable conditions, or dispersal flights triggered by population density or resource depletion.
2. **Environmental and Climatic Factors**
   - Temperature, wind patterns, and humidity levels can significantly influence the migration patterns of biological control agents. For example, many insects use wind currents to aid their long-distance travel, effectively expanding their range and influence over larger areas.
3. **Landscape Features**
   - Geographic features such as rivers, mountains, and roads can facilitate or hinder the movement of these agents. Understanding how landscape structure affects the movement patterns of biological control agents is crucial for optimizing their deployment.

Applications in Pest Management

Effective application of dispersion and migration knowledge can profoundly impact the management of pest populations, particularly in large-scale agricultural settings or integrated pest management (IPM) schemes.

1. **Strategic Release Programs**
   - Programs that strategically release biological control agents in patterns or locations that maximize their natural dispersion and migration behaviors can achieve more effective and enduring pest control. This approach ensures that agents can naturally spread throughout a pest-infested area without continuous human intervention.
2. **Augmentation of Natural Populations**
   - Enhancing the natural populations of control agents through supplemental releases that coincide with natural migration patterns can bolster the resilience and effectiveness of these populations, particularly in fragmented landscapes or where natural habitats have been altered.
3. **Barrier Management**
   - Managing landscape features that act as barriers to dispersion can enhance the spread of biological control agents. This might involve creating corridors that facilitate movement or modifying habitats to make them more conducive to the agents' survival and dispersal.

Benefits of Optimizing Dispersion and Migration

The optimization of dispersion and migration behaviors offers several benefits in the context of biological control:

1. **Enhanced Coverage**
   - By facilitating the spread of control agents across a wider area, pest populations can be managed more comprehensively, reducing the likelihood of pest resurgence or escape.
2. **Sustainability**
   - Dispersion and migration behaviors help establish self-sustaining populations of biological control agents, reducing the need for repeated interventions and the associated costs and labor.
3. **Resilience to Environmental Changes**
   - Dispersive and migratory behaviors allow biological control agents to adapt to environmental changes, maintaining their effectiveness even as conditions change due to climate variability or other factors.

Challenges and Future Directions

While dispersion and migration are advantageous, they also present challenges that need to be managed:

1. **Predictability**
    - The predictability of dispersion and migration patterns can be affected by environmental variability, making it challenging to consistently achieve optimal placement and timing of releases.
2. **Genetic Considerations**
    - The genetic basis of dispersal and migration behaviors needs careful consideration to avoid unintended consequences, such as the dilution of locally adapted traits or the potential for invasive spread.
3. **Integration with Other Control Methods**
    - Dispersion and migration need to be integrated with other pest management strategies to ensure comprehensive and effective control, particularly in diverse agricultural ecosystems.

iii. Adaptation of Predatory and Parasitic Insects

Adaptation in predatory and parasitic insects encompasses a wide range of evolutionary modifications that enhance their survival and efficacy within specific ecological niches. These adaptations can be morphological, physiological, or behavioral and are typically developed in response to environmental challenges, prey availability, and competitive interactions.

a. Adaptation in Biological Control Agents

Adaptation in biological terms refers to the changes organisms undergo to improve their chances of survival and reproduction in a given environment. For predatory and parasitic insects, such adaptations are crucial for navigating the complex dynamics of ecosystems where they must constantly find, subdue, and consume prey or successfully parasitize hosts.

1. **Morphological Adaptations**
    - These are physical changes in the structure of an organism that improve its functional capacity in the environment. For predatory and parasitic insects, morphological adaptations may include alterations in body size, shape, or color, and the development of specialized organs that aid in predation or parasitism.
2. **Physiological Adaptations**
    - These adaptations involve changes in internal functions, including metabolism, reproduction, and internal defense mechanisms. For instance, many parasitic insects have evolved physiological processes that allow them to suppress the immune responses of their hosts, enabling them to feed and develop unimpeded.

3. **Behavioral Adaptations**
   - Behavioral adaptations are changes in the way organisms act, which can include their feeding habits, mating rituals, and other aspects of their behavior. In predatory and parasitic insects, this might manifest as changes in how they hunt, evade predators, or select and utilize their hosts.

Key Examples of Adaptations in Predatory and Parasitic Insects

The adaptation strategies of these insects are diverse, each tailored to optimize their survival and predatory or parasitic capabilities within their environmental context.

1. **Camouflage and Mimicry**
   - **Camouflage**: This is a form of morphological adaptation where insects develop patterns and colors that blend into their surroundings. For example, the green praying mantis's coloration mirrors the green vegetation where it resides, making it nearly invisible to both its prey and potential predators.
   - **Mimicry**: Some predatory and parasitic insects mimic the appearance of other harmful or benign organisms to evade detection or to deceive their prey and predators. An example is some species of hoverflies that mimic the coloring and shape of wasps, deterring predators due to the perceived threat of stinging.
2. **Venom and Digestive Adaptations**
   - Many predatory insects have developed venom that paralyzes prey, allowing for easier consumption or provisioning for their larvae. Spider wasps, for instance, use their venom to paralyze spiders before laying an egg on them, providing a fresh food source for their emerging larvae.
   - Digestive adaptations in parasitic insects might include enzymes that can break down host tissues or counteract the host's defense chemicals, facilitating easier access to nutrients.
3. **Sensory Enhancements**
   - Predatory and parasitic insects often exhibit heightened sensory adaptations that allow them to detect prey or hosts from great distances or through barriers. For example, many wasps have highly developed olfactory senses that enable them to detect host insects through chemical signals emitted by the hosts or their habitats.

4. **Reproductive Strategies**
    - Adaptations related to reproduction are particularly notable in parasitic insects. Some, like the parasitoid wasps, can lay hundreds to thousands of eggs within a host, ensuring a high probability of progeny survival and continuation of their genetic lineage. This prolific reproduction strategy is closely tied to the availability of hosts and can be a key factor in the population control of various pest species.

Ecological and Practical Implications

The adaptations of predatory and parasitic insects are not just of academic interest but have practical implications for agriculture and pest management:

1. **Enhanced Biological Control**
    - By understanding and utilizing the natural adaptations of these insects, biocontrol practitioners can better plan and implement control strategies that are harmonious with natural ecological processes.
2. **Sustainability in Agriculture**
    - Leveraging these natural adaptations helps reduce reliance on chemical pesticides, promoting more sustainable agricultural practices that are less harmful to the environment and non-target species.
3. **Biodiversity Conservation**
    - Supporting the diverse adaptations of these insects contributes to the overall health and resilience of ecosystems. By maintaining populations of these adapted predators and parasites, we help preserve the intricate balance of our ecological systems.

b. Specialized Feeding Apparatus in Predatory and Parasitic Insects

In the vast and intricate world of predatory and parasitic insects, the evolution of specialized feeding apparatuses is a critical adaptation that enhances their ability to exploit specific prey or hosts effectively. These specialized structures vary widely among different species, each tailored to meet the unique demands of their feeding habits and ecological roles.

Specialized Feeding Apparatus

Specialized feeding apparatuses in insects are physical adaptations that have evolved specifically to improve feeding efficiency and reproductive success. These structures are incredibly diverse, ranging from the mandibles of predatory beetles to the ovipositors of parasitic wasps. Each adaptation not only reflects the ecological niche of the insect but also influences its interaction with other species within the ecosystem.

1. **Morphological Diversity**
   - **Predatory Adaptations**: In predatory insects, feeding structures are often designed to maximize the efficiency of capturing and subduing prey. For instance, the enlarged mandibles of ground beetles are adapted for crushing and slicing through the exoskeletons of their prey.
   - **Parasitic Adaptations**: Parasitic insects, such as wasps, have developed structures like ovipositors, which are used not only for laying eggs but also for delivering them into or onto the bodies of host organisms. This ensures that the developing larvae have direct access to the nourishment provided by the host's body.
2. **Physiological Enhancements**
   - Beyond mere structural adaptations, physiological enhancements complement these physical features, enabling insects to secrete digestive enzymes or venoms that immobilize prey or modulate host immune responses. These physiological traits are often closely integrated with the morphological characteristics of the feeding apparatus.

Evolutionary Drivers of Specialized Feeding Apparatus

The evolution of specialized feeding apparatuses in predatory and parasitic insects is driven by several key factors:

1. **Dietary Requirements**
   - The specific dietary needs of insects significantly influence the development of specialized feeding structures. Predators that consume hard-bodied insects may develop stronger, more robust mandibles or piercing mouthparts, whereas parasitic insects that need to breach the external surfaces of hosts evolve precise and often stealthy injection mechanisms.
2. **Environmental Pressures**
   - Environmental factors such as competition for food resources and predation pressure can drive the evolution of more effective and efficient feeding adaptations. Insects that can capitalize on a niche by exploiting specific prey types often have a competitive advantage, which can be enhanced by specialized feeding structures.
3. **Reproductive Success**
   - For parasitic insects, the ability to successfully place offspring in environments where they have access to sufficient resources is crucial.

Specialized ovipositors that can penetrate host tissues or reach hidden body cavities can dramatically increase reproductive success.

Examples of Specialized Feeding Apparatus

To better understand the diversity and functionality of these adaptations, here are some detailed examples:

1. **Beetle Mandibles**
    - Predatory beetles, such as the stag beetle, have evolved large, powerful mandibles that they use to grasp and crush their prey. These mandibles can vary significantly in size and shape, reflecting adaptations to specific types of prey or competitive interactions with other predators.
2. **Wasp Ovipositors**
    - The ovipositors of parasitic wasps are perhaps one of the most intricate examples of specialized feeding apparatus. These structures can be several times the length of the wasp's body and are used to deposit eggs directly into the living tissues of hosts. The ovipositor must be strong enough to penetrate the host's body but flexible enough to navigate to the optimal injection site.

Implications for Ecosystems and Biological Control

The presence of insects with specialized feeding apparatuses has significant implications for the ecosystems they inhabit and for strategies in biological control:

1. **Ecosystem Impact**
    - These insects play critical roles in regulating prey populations, contributing to the balance of ecosystems. Their specialized feeding strategies allow them to effectively control prey species that might otherwise become pests.
2. **Biological Control Applications**
    - Understanding these adaptations can aid in the development of targeted biological control programs. By selecting and utilizing insects with specific feeding apparatuses, pest populations can be managed more effectively and sustainably.
3. **Biodiversity and Ecological Interactions**
    - The specialization of feeding apparatuses often leads to complex ecological interactions, promoting biodiversity. These interactions can include predator-prey dynamics, competition, and mutualism, all of which contribute to the stability and resilience of ecosystems.

iv. Host-Seeking Behavior of Predatory and Parasitic Insects

Host-seeking behavior is a critical component in the life cycles of predatory and parasitic insects, encompassing the various strategies and mechanisms these organisms utilize to identify, locate, and exploit their prey or hosts. This behavior is not merely instinctual but often involves complex sensory integration and cognitive processes that significantly enhance the insects' survival and reproductive success. A thorough understanding of host-seeking behavior can profoundly improve the strategic application of these insects in biological control programs.

Chemical Cues

Chemical cues are among the most important tools predatory and parasitic insects use to locate their hosts or prey. These cues can come from a variety of sources and are detected through highly specialized sensory systems.

1. **Pheromones**: Insects often communicate through chemical signals, and pheromones play a significant role in this communication. Parasitoid wasps, for instance, are adept at detecting specific pheromones released by their host insects. These pheromones may be indicators of stress, mating readiness, or other biological functions that signal the presence and state of potential hosts.
2. **Plant Volatiles**: When plants are attacked by herbivorous pests, they can emit specific volatile organic compounds (VOCs) as a distress signal. These signals are not only a cry for help to attract natural predators of the pests but also serve to warn neighboring plants. Predatory and parasitic insects have evolved to recognize these signals as indicators of potential food sources. For example, certain wasps can detect these volatiles, leading them directly to plants where potential hosts are feeding.
3. **Other Chemical Traces**: In addition to pheromones and plant volatiles, insects can leave behind other chemical traces such as feces or excretory products that can attract predators or parasitoids keen on finding them.

Visual and Acoustic Cues

While chemical cues are crucial, many predatory and parasitic insects also rely heavily on visual and acoustic cues to find their prey or hosts.

1. **Visual Cues**: Vision plays a significant role in the host-seeking behavior of many predatory insects. Dragonflies, for example, have large, multifaceted eyes that provide them with a nearly 360-degree view of their surroundings, allowing them to spot potential prey from great distances. Their ability to hover and swiftly maneuver also aids in their predatory efficiency.

2. **Acoustic Cues**: Acoustic signals are vital for certain predatory species, especially in environments where visual cues may be limited. Predatory bats, not insects but important biological control agents for nocturnal pests, use echolocation to navigate and locate prey in the dark. This ability to emit sound waves and listen for their echoes as they bounce off objects allows bats to detect, track, and capture flying insect pests like moths.

Learning and Memory

The ability to learn from past experiences and improve future responses is a significant aspect of the host-seeking behavior of predatory and parasitic insects. This cognitive aspect of insect behavior is critical for adapting to dynamic environmental conditions and enhancing hunting or parasitic efficiency.

1. **Learning from Success**: Predatory and parasitic insects can learn to associate specific cues with successful feeding or parasitism events. For instance, wasps might learn that a particular plant species or color is often associated with the presence of their host insects. Over time, they will become more efficient at targeting these plants.
2. **Memory Retention**: The retention of successful strategies or locations is essential for repeated success in these insects' life cycles. Memory allows them to return to areas where they previously found hosts or prey, reducing the time and energy spent searching for food sources.
3. **Adaptive Learning**: Insects can also adapt their behavior based on environmental changes or previous unsuccessful attempts. If a usual strategy fails due to changes in prey behavior or environmental conditions, they can alter their approach, learning to overcome new challenges.

v. Field Collection and Preservation of Parasitoids and Predators

1. **Collection Areas:** Extensive field surveys are conducted in various districts of the Bundelkhand region to explore the presence of insect pests and their natural enemies such as parasitoids and predators. These surveys aim to cover a wide array of field crops, assessing the biodiversity and interactions within these agricultural ecosystems.
2. **Collection Equipment:** To facilitate the effective collection of insects, a comprehensive collection kit is essential. This kit includes:

- Scissors and forceps for delicate handling and extraction of specimens.
- Soft camel hair brushes for gently transferring insects without harm.

- Collection tubes, polythene bags, and paper bags for temporary containment.
- Specimen jars and tubes filled with 75-90% alcohol for preservation.
- Muslin cloth and rubber bands to secure containers.
- A hand lens for close examination, absorbent paper for moisture control, a field diary for record-keeping, and a camera for visual documentation of specimens and their habitats.

3. Collection of Insect Predators of Crop Plant and Weed Insect Pests

Insect predators are collected through two primary methods:

- **Field Collection:** Predators are directly collected from the field while they are actively engaging with or consuming their host pests.
- **Laboratory Rearing:** For some predators of aphids and whiteflies, immature stages are reared in controlled laboratory settings alongside their host pests to monitor their development and behaviors.

4. Collection of Insect Parasitoids of Crop Pests and Weeds

Similar to predators, insect parasitoids are collected from host-insect pests primarily through:

- **Laboratory Rearing:** Parasitoids are bred in laboratory conditions, where they are allowed to parasitize host pests. This controlled environment facilitates detailed observation and collection of various developmental stages.

5. Cataloguing of Field/Laboratory Data

Maintaining detailed records is crucial for the success of field and laboratory studies. Data cataloguing involves:

- **Use of Filling Cards:** Small cards (6 cm X 4 cm) are used to record vital information such as the host plant, host pest, types and numbers of parasites and predators collected, sex ratios, percentages of parasitism or predation, location details, altitude, and additional remarks.
- **Systematic Organization:** All data gathered from the field and laboratory are organized systematically in catalogues to facilitate easy access and reference, enhancing the utility of the collected data for research and application in pest management strategies.

Table: Common Predators of Crop Pests

S No	Order	Family	Characteristics and Behavior	Examples
1	*Odonata*		Larger-sized, aquatic immature stages (naiads), labium modified into a prehensile mask, adults feed during flight using basket-shaped legs	Dragon fly (*Anisoptera*), Damsel fly (*Zygoptera*)
2	*Dictyoptera*	*Mantidae*	Large, elongate, cryptic coloration, prehensile raptorial forelegs, highly predaceous	Praying mantis (*Mantis religiosa*)
3	*Hemiptera*	*Reduviidae*	Assassin bugs, blackish/brownish, short three-segmented proboscis, mostly predaceous and some blood-sucking	*Harpactor costalis* on red cotton bug *Dysdercus cingulatus*
		Pentatomidae	Shield shaped, 5-segmented antennae, predaceous on lepidopterous larvae	*Eucanthecona furcellata* on red hairy caterpillar and gram caterpillar
		Belostomatidae	Giant water bug, elongate oval, flattened with raptorial forelegs, feed on various aquatic insects	
		Miridae	Elongated, soft bodied, few species are predaceous	Green mirid bug (*Cyrotorhinus lividipennis*)
		Veliidae	Ripple bugs, aquatic, live on water surface, brown or black in color	*Microvelia atrolineata* on various insects in rice ecosystem
4	*Neuroptera*	*Myrmeleontidae*	Ant lions, larvae construct pit falls, feed by sucking the internal contents of prey	
		Chrysopidae	Aphid lions or green lace wings, green color, golden/ copper eyes, larvae are predaceous mainly on aphids and other small insects	
5	*Diptera*	*Asilidae*	Robber flies, elongate, dense hairs, long strong legs, piercing mouthparts, predaceous on a variety of insects	
		Syrphidae	Hover fly, brightly coloured, good pollinators, maggots feed on aphids by sucking their body fluids	
6	*Coleoptera*	*Coccinellidae*	Lady bird beetles, small, oval, convex, brightly colored, feed on soft-bodied insects like aphids and mealybugs	*Rodolia cardinalis* on cottony cushion scale, *Icerya purchasi*

		Carabidae	Ground beetles, dark, shiny, somewhat flattened, mostly feed on caterpillars	*Anthia sexguttata*, *Ophionea indica*
		Cicindelidae	Tiger beetles, very active, brightly colored, capture prey with sickle-shaped mandibles	*Cicindela spp.*
		Staphylinidae	Rove beetles	*Paederus fuscipes* feeds on rice leaf folder
7	*Hymenoptera*	*Vespidae*	Wasps, collect various insects, construct nests made of mud	
		Sphecidae	Digger wasps, construct mud nests, feed young ones with insect caterpillars	
		Formicidae	About half of the members are predaceous upon insects	

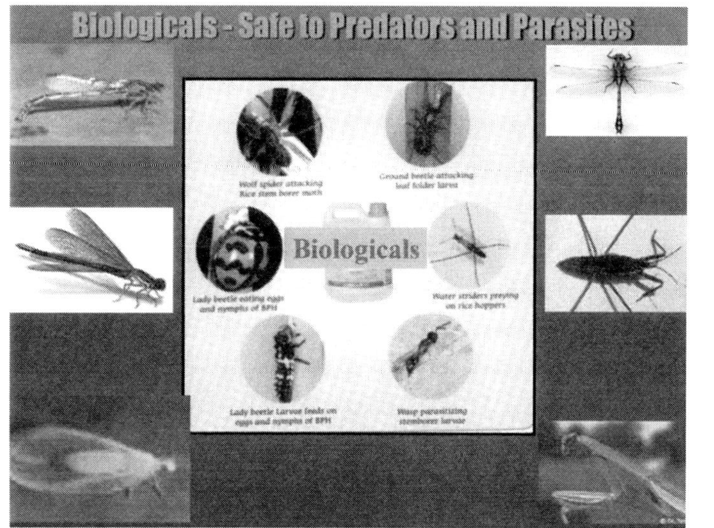

Table: Common Parasitoids of Crop Pests

S No	Order	Superfamily	Family	Characteristics and Behavior	Examples
1	*Hymeno ptera*	Ichneu-monoidea	*Ichneumonidae*	Long, filiform antennae; wings veined; abdomen long; larvae endo/ecto parasitic on insects and spiders; trochanter two segmented; two recurrent veins commonly, rarely one; large, slender insects with yellow, black, or reddish hues	*Eriborus trochanteratus* on coconut black headed caterpillar, *Opisina arenosella*
			Braconidae	Adults small, stout; abdomen as long as head and thorax combined; mostly endoparasitic on lepidopteran larvae; not more than one recurrent vein; less brightly colored compared to ichneumonids	*Bracon brevicornis* on *Opisina arenosella*; *Chelonus blackburni* on cotton bollworms
		Chalcidoidea	*Chalcididae*	Smallest parasitoids, gregarious; geniculate antennae; abdomen short or globular; wings without veins; hind femur enlarged and toothed; ovipositor straight and short; parasitic on Lepidoptera, Diptera, Coleoptera	*Brachymeria nephantidis* on *O. arenosella*
			Trichogramma-tidae	Minute insects (0.3-1.0 mm); three segmented tarsi; fore wings elongated with rows of microscopic hairs; hind wings reduced with hairs; mostly egg parasitoids	*Trichogramma chilonis* on various lepidopterous pests
			Eulophidae	Brilliant metallic colouring; males often have pectinate antennae; four segmented tarsi; mostly parasitic on aphids, scales, and some lepidopteran pupae	*Trichospilus pupivora*, *Tetrastichus israeli* on *O. arenosella*
		Bethyloidea	*Bethylidae*	Smaller than Ichneumonoidea but larger than Chalcidoidea; dark coloured, medium-sized wasps; females often wingless and ant-like; parasitic on Lepidoptera and Coleoptera; winged and wingless forms may occur in both sexes	*Parasierola (Goniozus) nephantidis* on *O. arenosella*
2	*Diptera*		*Tachinidae*	Large bristle flies; macrotype eggs attached to host's body with a glutinous secretion, microtype eggs laid on host plant for ingestion by host larvae; parasitic on various insects	*Sturmiopsis inferens* on sugarcane shoot borer, *Chilo infuscatellus*
3	*Lepidoptera*		*Epiricanidae*		*Epiricania melanoleuca*

Table: Common Entomo-Pathogens of Crop Pests

S No	Category	Type	Characteristics and Behavior	Examples
1	Entomo-pathogenic bacteria	*Bacillus thuringiensis*	Comprises a complex of bacterial subspecies; produces a parasporal body during sporulation containing highly toxic proteins in crystalline form; toxins are endotoxins, activated by proteolysis in the insect gut; destroys midgut epithelial cells, causing death.	*Bt kurstaki* (130-140 Kd) for Lepidoptera, *Bt kurstaki* HD1 (71 Kd) for Diptera and Lepidoptera, *Bt tenebrionides* (77-73 Kd) for Coleoptera, *Bt israeliensis* (125-145 Kd) for Diptera
2		*Bacillus popilliae*	Causes Scarab milky disease; disease results in an opaque white color in diseased larvae due to accumulation of sporulating bacteria in the hemolymph.	
3		*Bacillus sphaericus*	An aerobic, rod-shaped, gram variable bacterium; produces terminal/subterminal spherical spores; highly persistent in the environment; excellent control over Culex, Psorophora, and Mansonia mosquito larvae.	
4	Baculoviruses	Nuclear polyhedrosis virus (NPV)	Virus consists of proteinaceous polyhedral occlusion bodies embedding the virions or virus rods; causes insects to become dull in color, reduces feeding rate, and larvae to turn pinkish white on the ventral side with accumulation of polyhedra; in advanced stage, larvae become flaccid, skin becomes fragile, and eventually ruptures, typically hanging upside down from plants (tree top disease or Wipfelkrankeit Disease).	AcNPV in *Autographa californica*, LdNPV in *Lymantria dispar*, SlNPV in *Spodoptera litura*, HaNPV in *Helicoverpa armigera*, BmSNPV in *Bombyx mori*
5		Granulovirus	Similar to NPV but typically causes infections in specific hosts; virions are embedded in granular occlusion bodies.	TnGV in *Trichoplusia ni*, SiGV in *Sturmiopsis inferens*

6	Entomo-pathogenic fungi	White halo fungus	*Verticillium lecanii* causes infections in insects like the coffee green scale; mummifies and hardens the body of the scale insect, which is then covered with filamentous white hyphae. Infected scales are often found stuck to leaf veins with spores on the surface.	Infects *Coccus viridis*
7		Green muscardine fungus	*Metarhizium anisopliae* infects beetles and grubs such as the coconut rhinoceros beetle; mummifies the body which shrinks from its original 'C' shape and becomes dried to a hard structure covered with a dark olive green powdery mass consisting of spores.	Infects coconut rhinoceros beetle/grub
8		White muscardine fungus	*Beauveria bassiana* attacks various insects including silkworms and semiloopers; similar to the green muscardine fungus, it mummifies the host, shrinking and hardening the body which becomes covered with spores.	Attacks silkworm, castor semiloope

B. Role and Mode of Action of Insect Pathogenic Agents

Insect pathogenic agents such as nematodes, viruses, bacteria, fungi, and protozoa play crucial roles in the biological control of insect pests. Each group has unique characteristics and modes of action, which can be harnessed for integrated pest management (IPM) strategies. These mechanisms is essential for developing effective biocontrol programs.

1. Insect Pathogenic Nematodes

i. Role of Insect Pathogenic Nematodes

Insect pathogenic nematodes (IPNs) are crucial in biological control due to their ability to suppress various insect pest populations effectively. Their roles can be categorized into the following:

1. **Biological Control Agents:** IPNs, especially those from *Steinernema* and *Heterorhabditis* genera, are potent biological control agents due to their parasitic nature and ability to deliver bacteria that kill their hosts. Their primary targets include soil-dwelling insects like white grubs, root weevils, fungus gnats, and certain caterpillars that attack crops. These

nematodes complement other biological control agents by targeting pests that are challenging to control with conventional methods.

2. **Integrated Pest Management (IPM):** IPNs integrate well into IPM programs because they pose minimal risk to beneficial insects, are environmentally friendly, and can persist in the soil for extended periods. Their specificity enables selective pest control, reducing the adverse effects on non-target organisms and biodiversity.
3. **Reduction of Chemical Pesticide Use:** Their effectiveness in controlling insect pests allows for a reduction in chemical pesticide use. This leads to healthier ecosystems, reduced pesticide residues on crops, and less environmental contamination.
4. **Sustainable Agriculture:** IPNs provide farmers with a sustainable pest control method that aligns with organic and environmentally friendly agricultural practices. They are compatible with crop rotation, intercropping, and other sustainable farming methods.

ii. Mode of Action of Insect Pathogenic Nematodes

The mode of action of insect pathogenic nematodes involves a complex interplay between the nematode and its symbiotic bacteria, which work together to subdue and kill the host insect.

1. **Host Location and Penetration:** Infective Juveniles (IJs), the only free-living stage of IPNs, actively seek out insect hosts. They detect hosts by sensing carbon dioxide, temperature changes, and host excretions. Once a suitable host is found, the IJs penetrate the insect's body through natural openings such as the mouth, spiracles, or anus, or by directly burrowing through the insect's cuticle.
2. **Bacterial Release:** Inside the insect's hemocoel (body cavity), the nematodes release their symbiotic bacteria from their gut. These bacteria, *Xenorhabdus* (associated with *Steinernema*) or *Photorhabdus* (associated with *Heterorhabditis*), multiply rapidly in the insect's hemolymph (blood). The bacteria produce toxins that paralyze and kill the insect within 24-48 hours.
3. **Toxin Production and Host Death:** The bacterial toxins cause septicemia, killing the insect host quickly. The toxins impair the host's immune system and disrupt its neuromuscular functions, leading to paralysis. The bacteria also secrete enzymes that break down the insect's internal tissues, which the nematodes use as a nutrient source.
4. **Nematode Reproduction:** With the host immobilized and dead, the nematodes feed on decomposed host tissues and bacteria. They undergo

several molts, growing into adults that lay eggs within the host. The eggs hatch into juvenile nematodes, which develop into IJs inside the host cadaver. These new IJs then emerge from the decomposing insect and seek new hosts, thus continuing the life cycle.

iii. Epizootiology, symptomatology, and etiology of diseases caused by Insect Pathogenic Nematodes

a. Epizootiology of Insect Pathogenic Nematodes Insect pathogenic nematodes, notably from the families *Steinernematidae* and *Heterorhabditidae*, serve as crucial biological control agents targeting soil-dwelling insect pests. These nematodes operate within a complex life cycle intricately tied to symbiotic relationships with specific bacteria. The life cycle of these nematodes is characterized by several distinct phases that commence with the infective juvenile (IJ) stage. This stage, which is the only stage outside the host capable of surviving in the environment, actively searches for susceptible insect hosts. Infective juveniles are equipped with adaptations that allow them to penetrate host defenses through natural body openings or the cuticle.

Once a suitable host is found and penetrated, the nematode releases its symbiotic bacteria from its gut into the insect's hemolymph, initiating infection. This ability to breach insect defenses and introduce a secondary biological agent (the bacteria) is unique among biological control agents and crucial to their effectiveness.

Life Cycle of Pathogenic Nematodes

1. **Infective Juvenile Stage (IJ):** The infective juvenile (IJ) is the only stage capable of surviving outside the host, which is key to the nematode's ability to seek and infect new hosts. During this stage, they remain inactive but are primed to penetrate insect hosts using a combination of chemotactic and behavioral cues to locate them. The IJs penetrate through natural openings like the mouth, anus, or spiracles, or directly bore through the insect's cuticle. Once inside, the IJs navigate to the hemolymph (insect blood) and initiate the infection process.
2. **Release of Symbiotic Bacteria:** The primary mechanism of pathogenicity hinges on the nematode's symbiotic relationship with bacteria. Species of Steinernema release *Xenorhabdus* bacteria, while Heterorhabditis nematodes release *Photorhabdus*. Both bacteria rapidly multiply inside the insect host, secreting toxins that overwhelm the insect's immune defenses, ultimately leading to septicemia. These toxins immobilize and kill the insect host within 24-48 hours.

3. **Feeding and Reproduction:** After the bacteria start to kill the host, the nematodes begin feeding on the decomposing tissues and multiplying bacterial colonies. This results in rapid growth and reproduction. Steinernematids and heterorhabditids can undergo multiple generations within a single host, producing new IJs that eventually exit the cadaver.
4. **Dispersal:** As the host decomposes, the new generation of IJs disperses into the soil to locate new hosts, continuing the cycle.

Symbiosis with Bacteria

The mutualistic relationship between nematodes and their bacterial partners is fundamental to the pathogenicity of these biological control agents.

- **Bacterial Pathogenicity:** *Xenorhabdus* and *Photorhabdus* bacteria possess diverse virulence factors that attack the insect's immune system, quickly establishing a fatal infection.
- **Nutritional Support:** The bacteria also provide essential nutrients to the nematodes by decomposing host tissues, which the nematodes consume to develop and reproduce.
- **Secondary Metabolites:** Both bacterial genera produce secondary metabolites that deter other microbes from colonizing the insect cadaver, ensuring the nematodes and bacteria have an exclusive environment for development.

Host Range and Specificity

Insect pathogenic nematodes display a relatively broad host range, infecting various orders of insects, including Lepidoptera (moths and butterflies), Coleoptera (beetles), and Diptera (flies). They target soil-dwelling stages such as pupae, larvae, and occasionally adult insects that spend part of their life cycle in the soil. Host specificity varies by species:

- *Steinernema carpocapsae* is highly effective against lepidopteran pests like the codling moth (*Cydia pomonella*) and armyworms.
- *Heterorhabditis bacteriophora* targets beetle larvae, such as those of the Japanese beetle (*Popillia japonica*).

Environmental Adaptations

Pathogenic nematodes have adapted to diverse environments:

- **Steinernema:** Generally more resilient in colder climates and can persist for extended periods in the soil without a host.
- **Heterorhabditis:** Tend to be more active and aggressive, often seeking hosts in warmer climates.

Soil composition, moisture, and temperature greatly influence nematode activity. IJs are especially sensitive to desiccation, requiring moist soil to remain viable. They also demonstrate a preference for sandy or loamy soils where mobility is less restricted.

Application and Effectiveness

Pathogenic nematodes have been harnessed as biopesticides for a wide range of soil-dwelling pests. They are particularly effective in horticultural and agricultural systems that utilize integrated pest management (IPM) principles. Their ability to penetrate host defenses and kill pests quickly makes them ideal candidates for biocontrol, often reducing pest populations significantly within a single season.

Factors Affecting Efficacy

- **Host Density:** The higher the pest density, the more effective nematodes are because hosts are easier to locate.
- **Temperature:** Ideal temperatures range between 20-30°C, although certain nematode species thrive outside this range.
- **Soil Moisture:** Soil should be moist but not waterlogged.
- **UV Exposure:** IJs are highly susceptible to UV radiation and require shade or soil cover to remain viable.

Challenges and Future Prospects

Despite their utility, several challenges still exist in the broader application of insect pathogenic nematodes:

- **Storage and Formulation:** Developing stable formulations for mass production and extended shelf life remains a significant hurdle. IJs should remain viable from packaging to application.
- **Host Resistance:** Some insect pests are developing resistance to nematodes, necessitating the exploration of newer strains and species.

b. Symptomatology Induced by Nematode Infection

Insect pathogenic nematodes from the families Steinernematidae and Heterorhabditidae are among the most effective biological control agents against insect pests. Their mode of infection and the subsequent symptoms displayed by the infected insects are crucial for understanding the potential of these nematodes in pest management. These symptoms arise due to the nematodes' association with symbiotic bacteria like *Xenorhabdus* and *Photorhabdus*.

1. **Color Changes:** The first sign of infection is often a change in the color of the host insect. The pigments produced by the symbiotic bacteria, combined with the insect's immune response and internal decomposition, can manifest as color shifts, including darkening (black, brown, or gray), yellowing, and reddish tinges. The changes can be localized to specific body segments or spread uniformly throughout the body. These changes are helpful diagnostic features for identifying nematode infection.
2. **Behavioral Changes:** Nematode-infected insects often demonstrate marked behavioral shifts. These include:
   - **Reduced Mobility:** Infected insects become lethargic and exhibit significant reductions in activity due to the neurotoxic effects of bacterial metabolites.
   - **Reduced Feeding:** The toxic compounds impact the insect's ability to feed, leading to rapid cessation of feeding behavior.
   - **Abnormal Postures:** Nematode infection can cause twitching, staggering, or abnormal body positioning, such as hanging upside down, due to neuromuscular disruption.

These behavioral alterations make the host insect more susceptible to predation, reducing its fitness and increasing the chances of mortality.

3. **Physical Decay:** As the bacteria multiply within the insect hemolymph, the internal tissues begin to decay. This decay is accelerated by the enzymes and toxins secreted by both the bacteria and nematodes. Here are notable features of this physical degradation:
   - **Decomposition and Breakdown:** The insect's internal tissues liquefy, leading to significant body swelling and loss of structure. The internal organs become a nutrient-rich slurry, supporting the growth of nematodes and bacteria.
   - **Ruptured Body Cavity:** Insects at advanced stages of infection may rupture due to the buildup of gas and pressure inside the body. The ruptured cavity exposes a fluid or paste-like substance, representing the decayed tissues.
   - **Distinctive Odor:** Due to decomposition, infected insects often produce an unpleasant smell, further aiding in diagnosing nematode infection.
4. **Swollen and Mummified Insects**
   - **Swelling:** Hosts often swell due to internal tissue liquefaction and bacterial proliferation.

- **Mummification:** In cases where the bacterial growth is not sufficient to rupture the insect cuticle, the insect carcass may dry out and harden. This mummification leaves behind a rigid body encasing nematodes and bacteria.

5. **Nematode Emergence:** After completing their development within the host, nematodes seek new hosts. The infective juveniles (IJs) emerge en masse through the natural body openings or by rupturing the cuticle. This emergence signifies the last stage of infection and is a distinctive sign of nematode parasitism.

Implications for Pest Management: The characteristic symptoms induced by insect pathogenic nematodes highlight their efficiency and precision as biocontrol agents. The speed of immobilization, cessation of feeding, and rapid host death make these nematodes suitable for the prompt suppression of pest populations. They are particularly advantageous in integrated pest management strategies, where selective targeting is necessary.

Nematodes primarily infect soil-dwelling insect pests, making them essential agents in managing subterranean species such as rootworms, weevils, and borers. However, their natural compatibility with bacterial pathogens allows for rapid multiplication and effective colonization, making them viable against several other insect orders.

Factors Influencing Symptomatology: The variability in symptoms and their intensity depends on several factors:

- **Host Susceptibility:** The species, age, and health of the host significantly impact the manifestation of symptoms. Susceptible hosts tend to display rapid symptoms compared to partially resistant ones.
- **Environmental Factors:** Temperature, soil moisture, and pH can influence the activity of nematodes and their bacterial partners, altering the infection process.
- **Nematode Species:** Different nematode species show varying degrees of virulence, impacting the severity and speed of symptom development.

c. Etiology and Disease Progression of Nematode Infections

The successful infection of insect hosts by entomopathogenic nematodes is largely driven by the bacteria they harbor, specifically *Xenorhabdus spp.* and *Photorhabdus spp.* These bacteria are not only crucial for the pathogenicity of the nematodes but are also essential in the etiology and progression of the diseases caused by these nematode infections.

Etiological Agents: *Xenorhabdus spp.* and *Photorhabdus spp.*

Xenorhabdus and *Photorhabdus* bacteria are gram-negative, rod-shaped bacteria that live in the intestines of their respective nematode hosts. Each nematode species is typically associated with a specific bacterial species. These bacteria are released from the nematode's gut once it enters the insect host's hemolymph (blood equivalent in insects), where they begin their pathogenic action.

1. Initial Infection and Bacterial Release into the Hemolymph

The infection process begins when infective juvenile nematodes penetrate the insect host. Upon entering the hemolymph, the nematodes expel their bacterial symbionts. This release is critical as the hemolymph, rich in nutrients, provides an ideal environment for the bacteria to thrive.

2. Rapid Multiplication of Bacteria and Toxin Production

Once in the hemolymph, the bacteria multiply rapidly. Both *Xenorhabdus* and *Photorhabdus* produce a range of toxins and other virulence factors, such as proteases, lipases, and cytotoxins, which are instrumental in overcoming the host's immune defenses. These toxins disrupt the cellular integrity of the host by damaging cell membranes, inhibiting cellular enzymes, and inducing apoptosis (programmed cell death).

3. Immobilization and Death of the Host

The rapid proliferation of bacteria and the dissemination of toxins lead to systemic infection, which causes septicemia. The toxins lead to neuromuscular paralysis, and the overall immune suppression results in the rapid deterioration of the host's physiological functions. This phase is marked by visible symptoms such as lethargy, cessation of feeding, and eventual immobilization of the host, typically occurring within 24 to 48 hours post-infection.

4. Decomposition of Host Tissues

As the host insect dies, its internal tissues begin to decompose. The bacteria facilitate this process by secreting enzymes that break down host tissues, converting them into simpler compounds that can be easily assimilated. This decomposition process provides essential nutrients that promote the growth, reproduction, and development of both the bacteria and their nematode hosts. The nematodes consume the decomposed tissues and bacteria, growing into adults and reproducing within the host's body.

5. Completion of the Nematode Life Cycle

The final stages of the nematode life cycle are pivotal to its role as an effective biological control agent. Once the nematodes have established an infection within an insect host and successfully released their symbiotic bacteria, the bacteria's enzymatic action begins decomposing the host's internal tissues. During this phase, the nematodes thrive, using the liquefied tissues as a rich nutrient source. This feeding phase supports the growth and reproduction of the nematode population within the host cadaver.

Growth and Reproduction of the Nematodes

1. **Feeding and Maturation**: The infective juveniles (IJs), which are the initial stage that enters the host, feed on the decomposed tissue and bacteria. They develop into adults and reproduce within the insect's decomposing body cavity.
2. **Egg Production and Larval Development**: After mating, the female nematodes lay eggs within the cadaver. These eggs hatch into larvae that undergo successive molts, growing and feeding within the nutrient-rich cadaver until they reach maturity.
3. **Generation of New Infective Juveniles**: As the food resources within the insect are exhausted, the nematodes begin forming a new generation of infective juveniles. This stage is specially adapted for surviving outside the host and seeking new insect hosts. The IJs stop feeding, accumulate lipid reserves, and adopt a more resilient cuticle layer to withstand environmental challenges.

Emergence from the Cadaver and Host Seeking

Upon maturation, the IJs break out of the host cadaver and disperse into the surrounding soil. These nematodes are highly specialized and have evolved sophisticated host-seeking behaviors that allow them to detect potential insect hosts using chemical cues released by plants, insects, and the soil microbiome. Once they identify a new host, they repeat the infection cycle, propagating their population and ensuring continued control of soil-dwelling insect pests.

Implications for Biological Control

The unique life cycle of these nematodes makes them particularly well-suited for integrated pest management programs. Key advantages include:

- **Specificity**: Their infective juveniles target only specific insect pests, minimizing impacts on non-target species and the broader ecosystem.
- **Efficiency**: The rapid propagation of the nematode-bacteria partnership ensures quick and effective pest population suppression.
- **Persistence**: The IJs' resilience to environmental stressors allows them to persist in the soil for extended periods, providing ongoing protection against pest reinfestation.

2. Insect Pathogenic Viruses

i. Role of Insect Pathogenic Viruses

Insect pathogenic viruses, particularly those from the Baculoviridae family (Nuclear Polyhedrosis Viruses [NPVs] and Granulosis Viruses [GVs]), play a significant role in managing insect pest populations in agriculture, forestry, and horticulture. Their primary functions include:

1. **Biological Control Agents:** These viruses are naturally occurring pathogens that specifically target certain insect species. They can significantly reduce populations of harmful pests like caterpillars, sawflies, beetles, and aphids. The high specificity ensures minimal impact on non-target insects and helps maintain ecological balance.
2. **Compatibility with Integrated Pest Management (IPM):** Insect pathogenic viruses fit seamlessly into IPM programs. They can be used in combination with other biological control agents and compatible agricultural practices, reducing the reliance on chemical pesticides.
3. **Environmentally Safe Pest Control:** As naturally occurring organisms, these viruses do not pose significant risks to humans, animals, or the environment. They provide an eco-friendly alternative to chemical pesticides, which often have undesirable residues and adverse effects on non-target organisms.
4. **Resistance Management:** Because of their unique mode of action, insect pathogenic viruses help delay the development of resistance in pest populations, particularly when rotated or mixed with other control strategies.
5. **Sustainability:** The sustainability of using insect pathogenic viruses lies in their self-perpetuating nature. They replicate within infected insects, amplifying their numbers and spreading infection within pest populations. Their use aligns well with organic and sustainable farming practices.

ii. Mode of Action of Insect Pathogenic Viruses

The mode of action of insect pathogenic viruses involves a series of steps that exploit the biological processes of the host insect to establish and propagate the infection.

1. **Entry into the Host:** Insect pathogenic viruses are usually ingested by insect larvae through contaminated foliage or other food sources. The occlusion bodies (protein capsules containing virus particles) are dissolved in the alkaline environment of the insect midgut, releasing infectious virions.

2. **Invasion of Midgut Cells:** The freed virions penetrate the epithelial cells lining the insect's midgut. Inside these cells, the virus initiates replication, generating thousands of new viral particles that cause the cells to burst. These newly formed virions then spread to other organs through the hemolymph.
3. **Systemic Infection and Replication:** Once in the hemolymph, the virus invades other tissues like the fat body, tracheal matrix, and epidermis. The infected cells undergo multiple rounds of replication, producing large numbers of virus particles that accumulate in tissues.
4. **Symptoms and Host Death:** Infected insects often exhibit behavioral changes such as reduced feeding, abnormal movement, and slower development. As the infection progresses, the insect's body becomes swollen and discolored due to the accumulation of occlusion bodies. Eventually, the insect dies from the internal tissue breakdown caused by the viral replication.
5. **Virus Spread:** After death, the insect cadaver disintegrates, releasing occlusion bodies into the environment. These occlusion bodies contaminate plant surfaces, where they can be ingested by other insects and continue the infection cycle.

iii. Epizootiology, Symptomatology, and Etiology of Diseases Caused by Insect Pathogenic Viruses

Insect pathogenic viruses, particularly those in the Baculoviridae family (NPVs and GVs), are highly effective against specific insect pests. The epizootiology, symptomatology, and etiology of these viral diseases are crucial for their effective deployment in integrated pest management programs.

i. Epizootiology of Insect Pathogenic Viruses

Insect pathogenic viruses, particularly the baculoviruses (NPVs and GVs), play a crucial role in natural and applied biological pest control. The study of their occurrence, distribution, and spread within insect populations provides insights into how these pathogens contribute to regulating pest populations.

1. Natural Occurrence

- **Prevalence in Natural Populations:** Baculoviruses are naturally present in low levels in various insect populations. They specifically target Lepidoptera (caterpillars) and Coleoptera (beetles), among other groups. In healthy ecosystems, these viruses maintain their hosts at sub-lethal densities, keeping pest populations from reaching damaging levels.

- **Epizootic Events:** During specific environmental conditions and high host population density, these viruses can cause sudden outbreaks, known as epizootics. Epizootics drastically reduce pest populations, helping to prevent further crop damage.
- **Host-Specific Pathogens:** Different baculovirus strains have evolved to target particular insect species. For example, Spodoptera exigua NPV primarily affects the beet armyworm, while Cydia pomonella GV targets the codling moth. This specificity ensures minimal collateral impact on non-target species and biodiversity.

2. Transmission Mechanisms

- Horizontal Transmission:
  - **Primary Mode:** The most common method of viral transmission among insects is horizontal. This transmission occurs when insects ingest occlusion bodies present on contaminated surfaces or plants.
  - **Persistence of Occlusion Bodies:** Baculovirus occlusion bodies can persist in the environment for extended periods, providing an enduring reservoir of infection that can be reactivated by susceptible hosts.
  - **Surface Contamination:** Foliage and soil surfaces contaminated with the virus from previous outbreaks serve as infection sources for future generations.
- **Vertical Transmission**
  - **From Parent to Offspring:** Although less common than horizontal transmission, vertical transmission ensures the virus continues to infect future generations.
  - **Mechanism:** The virus is passed through eggs laid by infected adults, infecting developing larvae before they emerge.
- **Environmental Factors Influencing Transmission:**
  - **Climate:** Temperature and humidity significantly impact the survival of occlusion bodies. Warm, moist environments enhance the persistence of occlusion bodies, facilitating transmission, while extreme temperatures can inactivate them.
  - **Alternate Hosts:** Alternate hosts can act as reservoirs, harboring the virus even when primary hosts are absent. This extends the virus's ability to survive seasonal or environmental changes.
  - **Natural Predators and Parasitoids:** Predators or parasitoids that consume infected hosts can carry and disperse viral particles,

inadvertently aiding in the horizontal transmission between different insect populations.

3. Role of Host Density and Environmental Conditions in Epizootics

- **Density-Dependent Outbreaks:** Viral transmission is often density-dependent, meaning the likelihood of infection increases as the host population becomes denser. High host densities facilitate the rapid spread of pathogens due to increased contact rates.
- **Seasonal Factors:** Epizootic events can be seasonal, with high transmission rates often coinciding with specific weather patterns or the presence of susceptible life stages of the insect host. For instance, spring and early summer may offer optimal conditions for viral outbreaks as insect larvae proliferate.
- **Agricultural Practices:** Crop monocultures can lead to high host densities, increasing the risk of viral outbreaks. Conversely, diverse planting reduces susceptibility by interrupting pest population continuity and limiting transmission.

4. Implications for Biological Control

- **Integration in Pest Management:** Understanding the epizootiology of insect pathogenic viruses aids in their use as biopesticides. Viral formulations can be strategically applied during periods of high host density, exploiting natural transmission dynamics.
- **Monitoring and Surveillance:** Effective biological control requires constant monitoring to identify early signs of infection and assess the likelihood of epizootics. Surveillance helps optimize application timing for viral formulations.
- **Resistance Management:** Continuous exposure to viral pathogens can drive pest populations to develop resistance. Monitoring for resistance patterns allows for diversified control strategies, reducing reliance on a single pathogen.

ii. Symptomatology of Diseases Caused by Insect Pathogenic Viruses

Insect pathogenic viruses, particularly the baculoviruses (NPVs and GVs), are known for causing distinctive symptoms in their host insects. The severity and nature of the symptoms depend on the virus type, host species, and infection stage. The observable signs provide insight into how these viruses operate, revealing their destructive impact on insect pests.

1. Behavioral Changes

- **Feeding Reduction:** Insects infected with baculoviruses show a marked reduction in feeding activity due to the neurotoxic effects of viral toxins and the physiological changes inflicted by the virus. This reduced consumption leads to weight loss and underdevelopment, hampering the insect's ability to complete its life cycle.
- **Slowed Movement:** Infected insects display slowed movements or exhibit erratic and uncoordinated behavior, which makes them more susceptible to predation or environmental hazards. These behavioral changes occur due to neuromuscular disruptions caused by viral replication.
- **Developmental Delays:** Infection often leads to arrested development, preventing larvae from pupating or adults from reproducing. The virus can disrupt key hormonal pathways, delaying or stopping molting processes.

2. Color Changes

- **Milky Appearance:** Nucleopolyhedroviruses (NPVs) cause characteristic color changes due to the accumulation of viral occlusion bodies within the insect's hemolymph. This accumulation results in a milky or opaque appearance, particularly evident in the later stages of infection.
- **Color Shifts in Granuloviruses:** Granuloviruses (GVs) cause other color changes based on the host insect's natural pigmentation and the viral strain. For instance, Cydia pomonella GV (CpGV) leads to a darkened or brownish discoloration in the infected codling moth larvae.

3. Physical Deformities

- **Swollen Body:** Viral infection disrupts cellular functions and leads to extensive tissue breakdown. As a result, the host insect's body becomes swollen due to internal accumulation of viral occlusion bodies and cellular debris.
- **Deformed Structures:** The virus can interfere with normal growth patterns, leading to malformed wings, legs, and other body structures. This deformation hampers mobility and feeding, further weakening the insect's survival.

4. Tree-Top Disease

- **Behavioral Response:** Caterpillars infected with NPVs often exhibit a phenomenon known as "tree-top disease." Before death, the infected larvae instinctively climb to the top of plants or trees and die while clinging to branches. This phenomenon is due to the virus manipulating the host's behavior, ensuring its body ruptures and spreads viral particles from an elevated position.

- **Cadavers:** The infected larvae form conspicuous cadavers hanging from foliage, where they liquefy or disintegrate over time, releasing viral occlusion bodies into the environment. This vertical distribution increases the likelihood of contamination of new hosts below.

5. Disintegration and Release of Occlusion Bodies

- **Fragile Body:** The infected insect's body becomes increasingly fragile due to extensive tissue destruction by the virus. When the insect eventually dies, its body ruptures, releasing viral occlusion bodies into the environment.
- **Liquefaction:** NPVs cause the insect body to liquefy upon death, providing a means to disperse viral particles efficiently. This liquefied mass contaminates the surrounding foliage, soil, or water, ready to infect new hosts.

6. Variations Based on Virus Type and Host

- **Lepidopteran Infections:** Baculoviruses predominantly target lepidopteran insects, leading to significant tissue damage and rapid host death. Larvae often display classic tree-top disease symptoms and liquefaction.
- **Coleopteran Infections:** Coleopteran hosts infected by baculoviruses, particularly granuloviruses, show darkened or swollen abdomens due to internal damage. The infected larvae often burrow into the soil before dying, releasing viral particles underground.
- **Dipteran Infections:** While NPVs and GVs mainly infect Lepidoptera and Coleoptera, some strains can target Diptera, causing significant internal hemorrhaging and darkened, swollen bodies.

7. Molecular Basis of Symptomatology

- **Viral Gene Expression:** Specific baculovirus genes regulate the induction of characteristic symptoms. For example, certain genes control the production of viral occlusion bodies, while others manipulate host immune responses or alter feeding behavior.
- **Toxin Production:** The viruses produce toxins that disrupt host cellular processes and interfere with normal physiological functions, causing developmental delays, paralysis, and eventual death.

8. Host-Pathogen Interactions

- **Immune Evasion:** Baculoviruses have evolved strategies to evade the insect immune system, allowing them to replicate unhindered within host cells. This evasion prolongs infection and enhances viral propagation.
- **Host Resistance:** Insect hosts can develop resistance to viral infection over generations, leading to reduced symptom severity or delayed symptom onset. However, this resistance varies based on genetic factors and environmental influences.

iii. Etiology of Diseases Caused by Insect Pathogenic Viruses

Insect pathogenic viruses, primarily from the Baculoviridae family, play a significant role in managing insect pest populations. Understanding the etiology of the diseases they cause helps improve their application in integrated pest management programs.

Causative Agents

- **Baculoviridae Family:** This viral family primarily causes diseases in insects, particularly those of the orders Lepidoptera, Coleoptera, and Hymenoptera. The most notable pathogens include Nuclear Polyhedrosis Viruses (NPVs) and Granulosis Viruses (GVs). NPVs produce large occlusion bodies (OBs) that encapsulate multiple virions, while GVs produce smaller granular occlusion bodies that contain one or two virions.
- **Nuclear Polyhedrosis Viruses (NPVs):** Characterized by their polyhedral occlusion bodies, these viruses have genomes encoding dozens of proteins essential for infecting host cells. They mainly infect lepidopteran larvae, often leading to widespread epizootics in pest populations.
- **Granulosis Viruses (GVs):** These viruses form granular occlusion bodies with one or two virions. Their specificity targets particular insect species, such as Cydia pomonella GV, which infects the codling moth.

Infection Process

The infection process involves several stages, each vital for establishing a successful systemic infection.

1. **Ingestion**
    - **Exposure to Occlusion Bodies:** Insects ingest occlusion bodies present on contaminated foliage or surfaces. The OBs protect the virus against environmental conditions until the insect consumes them.

2. **Release of Virions**
    - **Alkaline Gut Environment:** In the insect's alkaline gut, the occlusion bodies dissolve, releasing virions (infective virus particles).
    - **Infection of Midgut Cells:** The virions infect the midgut epithelial cells, the primary site of initial viral replication.
3. **Replication and Spread**
    - **Rapid Replication:** The virus undergoes rapid replication within the midgut cells, producing new virions that eventually lyse the infected cells.
    - **Spread to Hemolymph:** The newly released virions enter the hemolymph (blood) and disseminate throughout the insect's body, spreading the infection to other tissues such as the fat bodies, tracheal matrix, and reproductive organs.
4. **Systemic Disease**
    - **Tissue Destruction:** The virus spreads to secondary tissues, where it replicates and destroys the cells. The extensive tissue destruction leads to organ failure, immune suppression, and host death.
    - **Septicemia:** The systemic infection results in septicemia due to bacterial proliferation, immune suppression, and toxin production. The insect becomes fragile, liquefies internally, and dies.

Host Specificity

Insect pathogenic viruses exhibit remarkable host specificity, with each species or strain infecting particular insect hosts. This specificity is crucial for targeted pest management.

- **Spodoptera litura NPV:** This virus targets the tobacco caterpillar and can devastate its populations.
- **Cydia pomonella GV:** This granulovirus specifically infects codling moth larvae and is widely used in orchards to control this pest.
- **Autographa californica NPV:** This virus infects a broad range of lepidopteran species and serves as a model organism for baculovirus research.

Virulence Factors

The pathogenicity of these viruses relies on a range of virulence factors that enhance their ability to establish infections, evade the host's immune system, and maximize virion production.

- **Viral Proteins**
  - **Enhancins:** These viral proteins degrade the peritrophic membrane in the insect gut, increasing the penetration of virions into midgut cells.
  - **Chitinases and Cathepsins:** These enzymes facilitate the breakdown of host tissues, aiding the release of virions and accelerating the infection process.
- **Immune Suppression**
  - **Apoptosis Inhibition:** Some viral genes inhibit host apoptosis, allowing viral replication to proceed without cellular suicide.
  - **Immune Evasion Proteins:** Viral proteins can bind to and neutralize host immune factors, preventing them from recognizing or responding to the viral infection.
- **Host Manipulation**
  - **Tree-Top Disease:** Some NPVs alter host behavior, prompting infected larvae to climb to the tops of plants before dying. This behavior enhances the spread of viral occlusion bodies across a wider area.

Environmental Factors and Disease Progression

- **Temperature and Humidity:** Ideal conditions for viral activity include moderate temperatures and high humidity. High temperatures can reduce virus viability, while low temperatures slow viral replication. High humidity promotes the persistence of occlusion bodies on foliage, facilitating their ingestion by susceptible hosts.
- **Rainfall and Soil Conditions:** Rain can wash occlusion bodies off foliage, reducing their availability to insect hosts. However, soil-dwelling pests can become infected by occlusion bodies persisting in the soil.

Evolution and Host Adaptation

- **Viral Evolution:** Insect pathogenic viruses evolve in response to host resistance mechanisms. They can gain mutations that enhance their infectivity, increase immune evasion capabilities, or expand their host range.
- **Host Adaptation:** Host insects can develop resistance to viral infections through genetic variation, reducing the virus's impact on their populations. However, such resistance may not protect against all viral

strains, maintaining a balance between pathogen virulence and host immunity.

Application in Integrated Pest Management

The unique properties of insect pathogenic viruses make them valuable tools in pest management programs.

- **Biopesticide Formulations:** Commercial formulations of NPVs and GVs provide targeted control of specific pests without harming beneficial insects or wildlife.
- **Compatibility with Other Agents:** Viral biopesticides can be used alongside other biological control agents or traditional insecticides, providing a multi-pronged approach to pest management.
- **Sustainability:** Their natural occurrence and host specificity make insect pathogenic viruses sustainable, environmentally friendly alternatives to chemical pesticides.

3. Insect Pathogenic Bacteria

i. Role of Insect Pathogenic Bacteria

Insect pathogenic bacteria are critical components of integrated pest management, targeting specific pest populations with remarkable precision and minimizing the impact on non-target species and the environment. They naturally occur in ecosystems and have evolved specialized mechanisms to infect, colonize, and kill their insect hosts. Key bacterial groups include Bacillus thuringiensis (Bt), Xenorhabdus spp., and Photorhabdus spp., which together provide a diverse arsenal of tools against various insect pests.

Key Pathogenic Bacteria

1. **Bacillus thuringiensis (Bt)**
    - **Overview and Diversity:** There are over 80 serovars of Bt, each possessing unique insecticidal toxins. Bt produces multiple classes of toxins, the most important being Cry proteins and Cyt proteins. Cry toxins are diverse, encompassing over 200 types, each showing specificity to different insect orders, including Lepidoptera (moths and butterflies), Coleoptera (beetles), and Diptera (flies and mosquitoes).
    - **Toxin Evolution and Specificity:** Cry toxins have evolved over millions of years to target specific pests with pinpoint accuracy. Their structure allows them to bind to unique receptors in the insect gut, ensuring that non-target organisms are largely unaffected. Cyt toxins act synergistically with Cry toxins and are particularly effective against Diptera.

- **Bt Products and Applications:** Bt-based bioinsecticides are applied as sprays in both conventional and organic agriculture. They can also be used to combat disease vectors like mosquitoes. Genetic modification of crops to produce Bt toxins internally offers season-long pest control, reducing the reliance on chemical pesticides.

2. **Xenorhabdus spp. and Photorhabdus spp.:**
   - **Symbiotic Associations:** These bacteria are integral to the life cycle of their nematode hosts, Steinernema and Heterorhabditis. Their mutualistic relationship involves the bacteria providing the nematodes with essential nutrients and immune suppression, while the nematodes ensure efficient delivery to insect hosts.
   - **Unique Toxin Production:** These bacteria produce a wide range of toxins and virulence factors:
   - **Tc Toxins:** These are tripartite complexes that disrupt insect cellular function by binding to the gut lining.
   - **Hemolysins:** They lyse insect blood cells, providing a nutrient-rich environment for bacterial growth.
   - **Antimicrobial Proteins:** Xenorhabdus and Photorhabdus produce bacteriocins that inhibit other microbes, reducing competition within the insect host.

ii. Mode of Action

1. **Toxin Production**
   - **Cry and Cyt Proteins (Bt):** Once inside the insect gut, Cry and Cyt toxins bind to specific midgut receptors, leading to pore formation, cell lysis, and leakage of gut contents. This disruption of digestion leads to starvation and eventual death. Secondary effects include septicemia due to opportunistic bacteria.
   - **Photorhabdus Toxins:** Photorhabdus produces multiple toxins, including Tc toxins and proteases that degrade the insect cuticle and hemolymph. These toxins target the insect nervous system and cytoskeleton, causing paralysis and systemic infection.
2. **Immune Suppression:**
   - **Bt Proteins:** Bt toxins are largely non-immunogenic, allowing the bacteria to evade the insect immune system. However, certain strains produce secondary metabolites that inhibit the host immune response.
   - **Nematode Symbionts:** Xenorhabdus and Photorhabdus suppress immune responses through multiple mechanisms, including disrupting

phenoloxidase pathways, inhibiting phagocytosis, and producing anti-inflammatory proteins.

3. **Nutrient Acquisition**
   - **Bt:** The degradation of the insect gut provides access to nutrients for bacterial growth and sporulation. Cry toxins increase gut permeability, allowing the bacteria to access the hemocoel.
   - **Symbiotic Bacteria:** Xenorhabdus and Photorhabdus rapidly replicate in the insect hemolymph, releasing enzymes that degrade host tissues and provide essential amino acids and lipids.

Application in Pest Management

1. **Bacillus thuringiensis (Bt)**
   - **Spray Formulations:** Bt formulations are often applied as sprays on various crops, offering effective control of pests like cabbage looper, corn earworm, and mosquito larvae.
   - **Transgenic Crops:** Genetically modified (GM) crops expressing Bt toxins have transformed pest management in crops like corn and cotton. These crops produce Cry proteins that target key pests like the corn borer and cotton bollworm, significantly reducing crop damage.
   - **Insect Resistance Management:** Bt crops are planted with refuge areas to prevent resistance development in target pests. Additionally, pyramid stacking (combining multiple Bt genes) enhances efficacy.
2. **Nematode-Symbiotic Bacteria**
   - **Soil Application:** Nematode-symbiotic bacteria are primarily applied via nematode carriers directly into the soil. They target soil-dwelling pests like root weevils, grubs, and rootworms.
   - **Specialized Strategies:** The bacteria can also be released through direct injection or nematode cadaver application to areas where pest populations are high.

Advantages and Challenges

Advantages

- **Specificity:** Bt and nematode-symbiotic bacteria target specific insect pests while preserving beneficial insects and reducing non-target impacts.
- **Efficacy:** These bacterial pathogens are potent even at low concentrations due to their complex virulence mechanisms and toxins.

- **Resistance Management:** Stacking multiple toxins and using genetically engineered strains provide significant advantages in managing pest resistance.

Challenges

- **Environmental Persistence:** Some Bt formulations degrade rapidly in sunlight, reducing their effectiveness in the field.
- **Resistance Development**: Insects can develop resistance through receptor modifications, prompting the need for novel strategies and regular monitoring.
- **Production Costs:** Scaling up production of nematode-symbiotic bacteria is costly due to the complexity of maintaining the nematode-bacteria relationship.

Future Directions

1. **Genetic Engineering**
   - **Enhanced Bt Toxins:** Developing novel Cry proteins through genetic engineering can overcome insect resistance and extend Bt's target range.
   - **Synthetic Biology:** Engineering Photorhabdus and Xenorhabdus to produce recombinant toxins or antimicrobial compounds will enhance their applicability in the field.
2. **Formulation Technologies**
   - **Nanoparticle Encapsulation:** Encapsulating Bt toxins in nanoparticles can protect them from environmental degradation, increasing field efficacy.
   - **Controlled Release Formulations:** These will allow for the gradual release of nematode-symbiotic bacteria, ensuring sustained pest control.
3. **Integrated Strategies**
   - **Biocontrol Combinations:** Combining Bt and nematode symbionts with predators, parasitoids, or fungal biocontrol agents can maximize pest suppression.
   - **Crop Management Practices:** Practices like crop rotation and intercropping can limit pest populations, reducing the pressure on bacterial pathogens.

4. **Microbial Ecology Studies**
    - **Nematode-Bacteria Dynamics:** Understanding the ecological interactions between nematode symbionts and other soil microbes will reveal ways to optimize their effectiveness.
    - **Bt Interactions:** Investigating how Bt interacts with natural enemies and beneficial microbes will identify synergies and improve pest management strategies.

iii. Epizootiology, Symptomatology, and Etiology of Diseases Caused by Insect Pathogenic Bacteria

1. Epizootiology of Insect Pathogenic Bacteria

Epizootiology, the study of disease occurrence and transmission patterns, is a crucial aspect of understanding how insect pathogenic bacteria regulate insect populations. This insight informs strategies for the biological control of insect pests. For insect pathogenic bacteria like *Bacillus thuringiensis* (Bt), *Xenorhabdus spp.*, and *Photorhabdus spp.*, the process of infection is heavily influenced by their natural distribution, environmental conditions, and insect behaviors.

1. Natural Occurrence and Distribution

- **Diverse Ecosystems**
    - Insect pathogenic bacteria exist in many environments. For example, Bt, known for its insecticidal properties, is ubiquitous in the soil and on vegetation. Its spores persist in the environment, giving it a broad ecological range.
    - The symbiotic bacteria *Xenorhabdus* and *Photorhabdus* inhabit soil-dwelling nematodes. These nematodes act as vectors, delivering the bacteria to insect hosts during infection. The bacteria, in return, rely on the nematodes for shelter and transport.
- **Habitat Specificity**
    - Different bacterial strains are adapted to specific ecosystems. For instance, certain Bt strains are found on plant surfaces, where they encounter insect larvae feeding on foliage.
    - *Xenorhabdus* and *Photorhabdus* symbionts are more common in soils, particularly in regions rich in organic matter that supports nematode activity.

2. Factors Influencing Disease Spread

- **Environmental Conditions:**
  - **Rainfall:** Rainfall patterns can enhance or reduce bacterial transmission. Rain can wash spores from foliage to the soil, while consistent moisture may promote bacterial growth.
  - **Temperature:** High temperatures generally enhance bacterial replication, but extreme heat can reduce bacterial survival. Cold temperatures slow down insect activity, reducing opportunities for infection.
  - **Humidity:** Humid environments promote bacterial proliferation, especially for Bt spores on leaf surfaces. Low humidity can dry out and kill bacterial spores.
- **Insect Behavior**
  - The feeding habits and movement patterns of insect pests greatly impact bacterial disease transmission.
  - Larval insects that aggregate in groups or feed indiscriminately are more susceptible to ingesting bacterial spores or nematodes carrying symbiotic bacteria.
  - Certain insect behaviors like cannibalism or social feeding increase the likelihood of disease spread among insect populations.
- **Natural Enemies and Alternate Hosts**
  - **Natural Enemies:** Predators and parasitoids often interact with bacterial pathogens, sometimes inadvertently assisting with bacterial dispersal. Predators eating infected prey or parasitoids feeding on infected hosts can become vectors.
  - **Alternate Hosts:** Insects that aren't primary targets for pathogenic bacteria can still act as reservoirs, maintaining bacterial populations when preferred hosts are scarce.

3. Transmission Mechanisms

- **Horizontal Transmission**
  - **Ingestion:** The most common transmission method is through ingestion. Insects feeding on contaminated foliage, soil, or plant residues ingest Bt spores or cells, which then germinate and establish infections.
  - **Nematode Vectoring:** For *Xenorhabdus* and *Photorhabdus*, horizontal transmission depends on nematode vectors. The nematodes

actively seek insect hosts, penetrate their body, and release the bacteria, which then infect the host.

- **Vertical Transmission**
    - Vertical transmission refers to the transfer of pathogens from parent to offspring. While less common, some bacterial strains can pass through insect eggs, ensuring continued infection across generations.
    - In the case of *Xenorhabdus* and *Photorhabdus*, infected nematode eggs harbor the bacteria, passing the pathogens to the next generation of nematodes.

Additional Considerations

- **Population Density**
    - Higher insect population densities increase the likelihood of bacterial transmission, as crowded insects more frequently interact with contaminated surfaces.
- **Host Susceptibility**
    - Susceptibility varies between insect species, depending on genetic factors, diet, and physiological defenses. Some insects have natural resistance to certain bacteria, while others are highly susceptible.
- **Epizootic Dynamics**
    - Epizootic outbreaks, where bacterial diseases rapidly spread through pest populations, are often cyclical. They can occur after a period of pest population buildup and may drastically reduce pest numbers.

2. Symptomatology of Diseases Caused by Insect Pathogenic Bacteria

Insect pathogenic bacteria induce a variety of symptoms that differ according to the bacterial species, the insect host, and the stage of infection.

The distinct symptoms associated with different bacterial pathogens is vital for diagnosing and effectively managing outbreaks.

1. Symptoms of *Bacillus thuringiensis* (Bt) Infection:

Bt is renowned for its effectiveness against a broad spectrum of insect pests, particularly lepidopteran and coleopteran larvae. Its mode of action involves the production of Cry toxins, which induce the following symptoms:

- **Feeding Suppression:**
    - Insects stop feeding shortly after ingesting Bt spores or toxins, leading to starvation. This feeding cessation occurs rapidly, often within hours, as the gut disruption impairs the insect's ability to process food.

- **Gut Perforation**
    - Cry toxins bind to specific receptors in the insect midgut, creating pores that perforate the gut lining. The result is a leakage of gut contents into the hemocoel, causing septicemia and toxemia as the insect's immune response is overwhelmed.
- **Color Change**
    - The immune response triggered by gut perforation leads to melanization, resulting in darkened larvae. The process involves encapsulation of damaged gut cells and toxins by melanin, a key immune compound in insects.
- **Death**
    - Insect death typically occurs within a few days of ingesting Bt toxins. The direct action of Cry toxins, compounded by starvation and septicemia, culminates in fatal disruption of the insect's physiological functions.

2. Symptoms of *Xenorhabdus* and *Photorhabdus* Infection

Xenorhabdus and *Photorhabdus* bacteria are symbionts of entomopathogenic nematodes that rely on these nematodes to infect their insect hosts. Once inside, they produce toxins and enzymes that overwhelm the host's immune system, leading to the following symptoms:

- **Septicemia**
    - Both bacterial genera induce septicemia following nematode infection. The bacteria multiply rapidly in the insect hemolymph, releasing potent toxins that paralyze and kill the host. The spread of bacteria throughout the insect's circulatory system results in rapid systemic infection.
- **Lethargy and Paralysis**
    - Infected insects display significantly reduced movement and eventually become paralyzed. Bacterial toxins disrupt the insect's neuromuscular functions, impairing motor coordination and halting escape responses.
- **Tissue Breakdown**
    - The bacteria secrete proteases and other enzymes that degrade insect tissues, softening the body and making it prone to rupture. The cadaver becomes a nutrient-rich environment where bacteria and nematodes can grow and reproduce.

- **Bioluminescence**
  - Some *Photorhabdus* species are bioluminescent, illuminating the cadavers of infected insects. This unusual trait has been studied for its ecological and evolutionary significance, as it may aid nematode dispersal or signal to potential predators.

3. Additional Symptomatology Observed in Insect Pathogenic Bacteria:

- **Behavioral Changes**
  - Many insect pathogenic bacteria induce behavioral changes in their hosts, such as disorientation, uncoordinated movement, and compulsive climbing behaviors. For example, some infected larvae climb higher on plants before succumbing, facilitating the dispersal of pathogens.
- **Odor and Decomposition:**
  - Decomposing cadavers often emit foul odors due to bacterial decomposition. The rapid breakdown of tissues is accompanied by gas buildup and leakage of fluids, which helps spread pathogens to other insects.
- **Changes in Developmental Progression**
  - Some infections may alter an insect's developmental timeline, causing developmental delays or accelerated aging. For instance, infected larvae might enter pupation prematurely or delay metamorphosis as a stress response.

3. Etiology of Bacterial Diseases in Insects

The etiology of bacterial diseases in insects is rooted in the biology and interactions of the causative pathogens and their insect hosts. Understanding the infection mechanisms and disease progression provides valuable insights for leveraging these bacteria in biological control programs. Here is an in-depth exploration of the etiology of *Bacillus thuringiensis* (Bt) and nematode-symbiotic bacteria (*Xenorhabdus* and *Photorhabdus*), highlighting the processes that underpin their pathogenicity:

1. *Bacillus thuringiensis* (Bt)

Causative Agents

- *Bt* is a spore-forming bacterium that infects insects primarily through its production of proteinaceous crystalline toxins known as Cry and Cyt proteins. These proteins are produced during sporulation and are key to Bt's insecticidal properties.

- Cry and Cyt proteins target specific insect groups due to their affinity for midgut receptors, resulting in a selective infection mechanism. Different strains of *Bt* produce distinct Cry proteins that exhibit varying host specificity.

Infection Mechanism

1. **Ingestion**
   - Insects ingest Bt spores and crystalline toxins from contaminated foliage. This contamination occurs when Bt formulations or spores are applied to crops, or when insects come into contact with naturally occurring Bt strains.
2. **Activation and Binding**
   - The alkaline conditions of the insect midgut dissolve the crystalline toxins, releasing active Cry and Cyt proteins. These proteins bind to specific receptors on the surface of midgut epithelial cells.
3. **Pore Formation and Gut Leakage**
   - After binding, the toxins undergo conformational changes that facilitate pore formation in the midgut cells. The pores cause osmotic imbalance and cell lysis, leading to leakage of gut contents into the hemocoel.
4. **Septicemia**
   - The disrupted gut lining allows opportunistic bacteria to enter the insect hemocoel, causing septicemia. The combined action of septicemia and Bt toxins results in host death within days.

2. *Xenorhabdus* and *Photorhabdus*

Causative Agents

- *Xenorhabdus* and *Photorhabdus* are symbiotic bacteria associated with entomopathogenic nematodes from the genera *Steinernema* and *Heterorhabditis*, respectively.
- These bacteria live in specialized compartments within the nematode's gut and rely on the nematode to invade insect hosts.

Infection Mechanism

1. **Nematode Invasion**
   - Infective juveniles (IJs) of nematodes actively seek out insect hosts, gaining entry through natural openings (mouth, anus, spiracles) or directly penetrating the cuticle.

- Upon penetration, the nematodes release their symbiotic bacteria into the insect hemolymph. The bacteria are protected in the nematode gut until the insect's immune defenses are breached.

2. Bacterial Multiplication

- Once released, *Xenorhabdus* and *Photorhabdus* multiply rapidly in the hemolymph, producing a range of toxins, proteases, and enzymes that degrade insect tissues and suppress immune responses.
- These toxins disrupt hemocyte function and inhibit melanization, allowing the bacteria to spread systemically and overwhelm the host's immune defenses.

3. Nutrient Acquisition

- The bacteria secrete proteases that digest the insect's internal organs, providing nutrients for both themselves and their nematode partners.
- This nutrient-rich environment supports nematode reproduction, allowing multiple generations to develop within the host cadaver.

Virulence Factors and Immune Suppression:

- *Xenorhabdus* and *Photorhabdus* produce a variety of virulence factors that contribute to their pathogenicity, including toxins that target hemocytes, proteases that degrade host proteins, and siderophores that scavenge iron.
- The bacteria suppress the insect immune system by inhibiting melanization and hemocyte aggregation, preventing encapsulation and phagocytosis of nematode juveniles.

Ecological and Evolutionary Context:

- *Xenorhabdus* and *Photorhabdus* are highly co-evolved with their nematode hosts, resulting in an efficient system for targeting insect pests.
- This symbiotic relationship enables nematodes to exploit a broader range of hosts than they could on their own, while providing bacteria with a consistent mode of transmission.

4. Insect Pathogenic Fungi

i. Role and Mode of Action of Insect Pathogenic Fungi

Insect pathogenic fungi play a vital role in regulating insect populations, naturally controlling pest outbreaks, and serving as foundational agents in integrated pest management (IPM) strategies. Their diverse mechanisms of infection, coupled with their ability to target specific insect hosts, make them invaluable tools for biological pest control.

1. **Overview and Ecological Role:** Insect pathogenic fungi, also known as entomopathogenic fungi, encompass a wide variety of fungal species that can infect and kill insects. They belong to multiple genera, including Beauveria, Metarhizium, Isaria (formerly Paecilomyces), Lecanicillium, and others. Their ecological role is crucial for the following reasons:
   - **Population Regulation:** They help regulate insect populations naturally by inducing epidemics or epizootics that drastically reduce pest densities, thereby maintaining a balance in ecosystems.
   - **Integrated Pest Management (IPM):** These fungi are increasingly used in IPM programs because they can be cultivated and formulated into commercial biopesticides. Their selective pathogenicity, reduced environmental impact, and compatibility with other control measures make them effective tools in pest management.
   - **Host Specificity:** Some species have broad host ranges, while others are highly specific to particular insect families, orders, or even genera. This specificity minimizes the risk of collateral damage to non-target organisms.
   - **Environmental Persistence:** These fungi can persist in the soil or on plant surfaces, often forming long-lived spores that remain viable for years, thereby providing a lasting impact.

2. Mode of Action

The mode of action of insect pathogenic fungi involves several distinct stages, from initial adhesion to spore germination, cuticle penetration, and systemic colonization of the host:

a. **Spore Adhesion**
   - **Attachment to Cuticle:** The process begins with the attachment of fungal spores, called conidia, to the insect cuticle (exoskeleton). This attachment is facilitated by specialized surface structures and proteins that recognize the cuticle composition.
   - **Enzymatic Secretion:** The spores secrete enzymes, such as cuticle-degrading proteases and lipases, that dissolve the insect's protective waxy layer, increasing spore adhesion.

b. **Germination and Penetration**
   - **Germ Tube Formation:** Upon attachment, the conidia germinate, producing germ tubes that extend toward the cuticle surface.

- **Appressorium Formation:** The germ tubes form specialized structures known as appressoria, which adhere tightly to the cuticle and generate immense turgor pressure. This pressure, coupled with enzymatic degradation, allows the appressorium to pierce the cuticle.

c. **Colonization and Immune Evasion**

  - **Hyphal Penetration:** After cuticle penetration, the fungal hyphae invade the insect's internal tissues, including the hemocoel (body cavity).
  - **Immune Suppression:** The fungi release secondary metabolites and toxins that suppress the insect's immune system, inhibiting the activity of hemocytes and preventing the production of antimicrobial peptides. This enables the fungi to colonize the hemocoel unchallenged.

d. **Systemic Infection and Toxin Production**

- **Systemic Hyphal Growth:** The fungi proliferate within the insect, spreading throughout the body and consuming hemolymph (blood), internal organs, and other tissues.
- **Mycotoxins:** Some fungi, such as *Beauveria bassiana* and *Metarhizium anisopliae*, produce mycotoxins that cause septicemia, leading to rapid paralysis and death.

e. **Host Death and Sporulation**

  - **Host Death:** Eventually, the fungal infection overwhelms the host, leading to death through a combination of nutrient depletion, systemic infection, and septicemia.
  - **Conidiogenesis and Sporulation:** The fungal hyphae then grow outward, breaching the cuticle to cover the host cadaver. They produce conidiophores that release conidia into the environment, restarting the infection cycle.

3. Notable Genera and Species

The diversity of insect pathogenic fungi is vast, but some genera stand out due to their effectiveness and application potential:

a. *Beauveria* **Species**

  - ***Beauveria bassiana*:** One of the most widely studied fungi, it is highly effective against beetles, aphids, and caterpillars. It is used in several commercial biopesticides due to its broad-spectrum activity.
  - ***Beauveria brongniartii*:** Commonly used for controlling the European cockchafer and other soil-dwelling insects.

b. ***Metarhizium*** Species

- ***Metarhizium anisopliae***: Effective against termites, grasshoppers, and many beetles. It forms part of commercial biopesticide formulations.
- ***Metarhizium brunneum***: Commonly used in IPM strategies against aphids, thrips, and whiteflies.

c. ***Isaria*** (formerly ***Paecilomyces***) Species

- ***Isaria fumosorosea***: Notable for its efficacy against whiteflies, aphids, and other soft-bodied insects.

d. *Lecanicillium* Species

- ***Lecanicillium lecanii***: Useful in greenhouses for controlling aphids and whiteflies.

4. Applications and Challenges in Pest Management

- **Formulation and Field Application**
  - These fungi are formulated as wettable powders, emulsifiable concentrates, or granules to facilitate easy application in different environments.
  - Careful calibration of spore concentrations is necessary to ensure effectiveness while minimizing unintended effects on non-target organisms.
- **Environmental Factors**
  - Environmental factors such as humidity, temperature, and UV radiation significantly influence fungal efficacy. High humidity and moderate temperatures are generally conducive to fungal infection, while UV radiation can degrade conidia rapidly.
- **Host Resistance**
  - Insect resistance to fungal pathogens may emerge through altered behavior (e.g., avoiding contaminated surfaces) or physiological changes (e.g., thicker cuticles or immune priming).
- **Compatibility with Other Control Methods**
  - Insect pathogenic fungi can be integrated with chemical, cultural, and other biological control measures to enhance pest management. However, the use of broad-spectrum fungicides should be minimized to prevent non-target impacts on beneficial fungi.

ii. Epizootiology, Symptomatology, and Etiology of Diseases Caused by Insect Pathogenic Fungi

Insect pathogenic fungi, also called entomopathogenic fungi, are vital components of ecosystems and integrated pest management (IPM) programs due to their ability to infect and control insect pests naturally. The dynamics of fungal diseases in insect populations are complex, and understanding their epizootiology, symptomatology, and etiology provides insights that can optimize their use in pest management.

1. Epizootiology of Insect Pathogenic Fungi

Epizootiology, which studies the distribution, occurrence, and spread of diseases in insect populations, provides valuable insights into how insect pathogenic fungi interact with their hosts and the environment.

a. Natural Occurrence and Distribution

- **Fungal Species:** Notable fungal genera include *Beauveria*, *Metarhizium*, *Isaria*, and *Lecanicillium*, all of which occur naturally in soil and on plant surfaces.
- **Host Spectrum:** While some fungal species exhibit a broad host range, others specifically target certain insect groups. For instance, *Beauveria bassiana* has a wide range of hosts, whereas *Lecanicillium lecanii* primarily targets aphids and whiteflies.
- **Persistence:** Fungal conidia can persist in the soil and on plant surfaces for prolonged periods, often in dormant states, ensuring long-term impact on host populations.

b. Factors Influencing Disease Spread

- **Environmental Conditions:** Temperature, humidity, and UV radiation influence fungal survival, germination, and infection rates. High humidity and moderate temperatures typically favor fungal infections, while UV radiation can degrade conidia.
- **Host Behavior:** The feeding habits and mobility of insect hosts can influence disease spread. Insect populations with higher densities or migratory patterns can lead to greater disease dissemination.
- **Natural Enemies and Alternate Hosts:** Predators and parasitoids can aid disease spread by preying on infected insects or directly facilitating fungal infection. Alternate hosts also maintain fungal populations during low pest activity periods.

c. **Transmission Mechanisms**

- **Horizontal Transmission:** This is the primary mechanism, where conidia attach to and penetrate the cuticle of susceptible insects, ultimately killing them.
- **Vertical Transmission:** Some fungi can also be vertically transmitted via infected eggs or through maternal contact.

2. Symptomatology of Diseases Caused by Insect Pathogenic Fungi

The symptoms caused by insect pathogenic fungi are varied and depend on the type of fungi, insect host species, and infection stage.

a. **Behavioral Changes**

- **Feeding Suppression:** Infected insects often reduce or stop feeding shortly after infection, leading to starvation.
- **Lethargy and Paralysis:** As fungal toxins disrupt neuromuscular function, infected insects become lethargic and may eventually become paralyzed.

b. **Physical Symptoms**

- **Color Changes:** Fungal infection often leads to discoloration of the host cuticle. Insects infected with *Metarhizium* species may develop a greenish hue, while those infected by *Beauveria* often appear white.
- **Mycelial Growth:** The fungus grows internally and externally on the insect, eventually covering it with visible mycelial filaments or conidiophores that produce conidia.
- **Disintegration:** As the fungus consumes internal tissues, the insect body weakens and may rupture, releasing new spores into the environment.

c. **Death and Sporulation**

- **Host Death:** The infected insect typically dies within days to weeks, depending on the fungal species and environmental conditions.
- **Sporulation:** Fungal hyphae grow outwards, covering the cadaver and releasing conidia to restart the infection cycle.

3. Etiology of Diseases Caused by Insect Pathogenic Fungi

The etiology of diseases involves understanding the causative agents, their infection mechanisms, and the factors influencing disease progression.

a. **Causative Agents**

- **Beauveria:** *Beauveria bassiana* and *Beauveria brongniartii* are among the most studied species, effective against a wide range of insect pests.

- **Metarhizium:** *Metarhizium anisopliae* and *Metarhizium brunneum* are effective against many soil-dwelling insects and are often applied in IPM programs.
- **Isaria (formerly *Paecilomyces*):** *Isaria fumosorosea* and *Isaria farinosa* are notable for their activity against whiteflies and aphids.
- **Lecanicillium:** *Lecanicillium lecanii* and related species specifically target aphids and thrips.

b. Infection Mechanisms

- **Attachment and Germination**
  - **Spore Adhesion:** The conidia attach to the insect cuticle through specialized proteins and enzymes that dissolve the protective waxy layer.
  - **Germ Tube Formation:** Once adhered, the spores germinate and produce germ tubes that extend toward the cuticle.
- **Cuticle Penetration and Colonization:**
  - **Appressoria Formation:** Germ tubes form appressoria, specialized structures that generate turgor pressure to pierce the cuticle.
  - **Hyphal Penetration:** After cuticle penetration, the hyphae spread throughout the insect's internal tissues.
- **Systemic Growth and Toxin Production:**
  - **Systemic Infection:** Fungal hyphae proliferate throughout the insect, colonizing the hemocoel and other tissues.
  - **Toxins and Enzymes:** Some fungi release mycotoxins that cause paralysis and septicemia, while others produce enzymes that degrade tissues.

c. Host Defense and Disease Progression:

- **Immune Suppression:** Fungal toxins suppress the insect immune response by inhibiting hemocyte activity and antimicrobial peptide production.
- **Nutrient Depletion:** The fungi consume hemolymph and internal organs, leading to nutrient depletion and death.
- **Septicemia and Tissue Breakdown:** The systemic infection, combined with immune suppression and mycotoxin production, causes septicemia and tissue degradation, ultimately killing the insect.

5. Insect Pathogenic Protozoa

i. Role of Insect Pathogenic Protozoa

A. Natural Pest Population Regulation

- **Ecological Balance:** Insect pathogenic protozoa maintain natural ecological balance by infecting and regulating insect populations. These protozoa prevent pests like locusts, beetles, and caterpillars from growing uncontrollably and damaging crops or other vegetation. Their presence ensures that pests don't exceed natural thresholds, allowing native plants and beneficial insects to thrive.
- **Adaptability:** Protozoa demonstrate remarkable adaptability to diverse environmental conditions, which helps them manage pests in varying climates and regions. Species like *Nosema locustae* can withstand temperature fluctuations and changes in humidity, ensuring consistent biological control. This adaptability enables protozoa to offer protection in both natural ecosystems and agricultural environments.
- **Epizootic Outbreaks:** Protozoa can cause significant disease outbreaks, or epizootics, under favorable environmental conditions. For instance, *Nosema locustae* often triggers widespread infections in grasshoppers and locusts, reducing their populations dramatically. These outbreaks are crucial for controlling pests, particularly during years of high pest density, minimizing the need for chemical insecticides.
- **Host-Specificity:** Most protozoan pathogens exhibit high host-specificity, meaning they target specific insect groups while leaving non-target species unaffected. This specificity ensures that beneficial insects, birds, and mammals remain unharmed. *Vavraia culicis* exclusively targets mosquito larvae, while *Nosema locustae* primarily infects grasshoppers. Such specificity helps maintain biodiversity while protecting crops.

B. Use in Biological Control

- **Commercial Biopesticides:** Some protozoan pathogens, like *Nosema locustae*, have been developed into commercial biopesticides. These biopesticides are applied strategically to manage pests, particularly in areas where chemical control is less effective or undesirable. They offer effective pest control solutions for organic farming or in areas near sensitive ecosystems.
- **Integration in IPM Programs:** Protozoa integrate seamlessly into Integrated Pest Management (IPM) programs because they complement

other biological control strategies like predators and parasitoids. Their compatibility with other biological agents ensures long-term and sustainable pest control. When combined with insect pathogenic fungi or bacteria, they offer comprehensive pest management without causing resistance issues commonly seen with chemical insecticides.

- **Environmental Persistence:** Protozoa often persist in the environment once introduced, ensuring sustained pest control. Their spores or cysts can withstand extreme temperatures or humidity levels and remain viable for long periods, even through winter. This persistence reduces the frequency of application required and ensures continued control of pests.

ii. Mode of Action of Insect Pathogenic Protozoa

A. Infection Mechanisms

- **Ingestion:** Insect pathogenic protozoa often infect their hosts through ingestion. Spores or cysts contaminate foliage, soil, or insect feces and are consumed during feeding. For example, *Nosema spp.* spores are ingested by grasshoppers from contaminated leaves. Once in the host gut, these spores germinate and infect intestinal cells.
- **Direct Inoculation:** Some protozoa can enter hosts directly via wounds or through parasitic vectors. For instance, *Leidyana migrator* attaches to the cuticle or is inoculated by parasitic mites, penetrating the host's body through damaged tissue.

B. Lifecycle and Pathogenicity

- **Microsporidia**
  - **Spore Germination:** Microsporidian spores germinate in the alkaline gut environment, releasing sporoplasm that directly injects into host cells via a specialized polar tube. This injection ensures that the infective stage reaches the host's internal tissues, facilitating rapid infection.
  - **Intracellular Multiplication:** Sporoplasm replicates intracellularly, exploiting the host's cellular machinery. This unchecked replication leads to the destruction of infected cells, weakening the insect. Over time, this results in lethargy, weight loss, and failure to molt.
  - **Spore Formation:** Once intracellular replication is complete, new spores form within infected cells, eventually causing cell rupture. The spores disseminate through the hemolymph or digestive tract, allowing infection to spread. These spores pass out of the insect via feces or when cadavers disintegrate, leading to horizontal transmission.

- **Gregorine Protozoa**
  - Ingestion and Infection: Gregarine protozoa rely on ingestion for infection. Insects consume infective spores or oocysts from contaminated food or water. Once in the gut, the spores hatch, releasing motile sporozoites that attach to the gut lining.
  - Attachment and Growth: Sporozoites develop into trophozoites, which mature while attached to gut epithelial cells. These trophozoites then detach and move through the digestive tract.
  - Reproduction and Transmission: Trophozoites undergo sexual reproduction to produce new infective spores, which are then excreted through insect feces, perpetuating the infection cycle.
- **Ciliates**
  - **Attachment and Feeding:** Ciliates attach to insect cuticles or invade internal tissues. Their hair-like cilia help them navigate the host's tissues and attach to cells. Once attached, they feed directly on the host's body fluids.
  - **Reproduction and Transmission:** They reproduce via binary fission or conjugation, then exit the host through feces or body fluids. These cysts infect new hosts via contaminated food or water.
- **Apicomplexa**
  - **Sporozoite Invasion:** Apicomplexan protozoa, like *Plasmodium*, infect insects via sporozoites. Insects ingest contaminated food or water, releasing sporozoites that infect the gut lining.
  - **Host Cell Invasion:** Sporozoites enter host cells, where they rapidly multiply and disrupt cellular integrity. These infected cells burst, releasing new infective stages.
  - **Systemic Pathogenicity:** Apicomplexa often cause systemic infections, damaging internal organs. They impair vital functions like digestion and respiration, eventually killing the host.

iii. Epizootiology, Symptomatology, and Etiology of Diseases Caused by Insect Pathogenic Protozoa

Insect pathogenic protozoa are a diverse group of microorganisms that can cause diseases in a wide range of insect species. Understanding the epizootiology (the study of the patterns and determinants of disease in non-human populations), symptomatology (the clinical manifestations of disease), and etiology (the cause of disease) of these protozoan-induced diseases is essential for effective pest management and control.

1. Epizootiology

a. **Host Range:** Insect pathogenic protozoa display varying degrees of host specificity, with some species targeting specific insect groups while others have a broader host range. For example, Nosema spp. primarily infect grasshoppers and locusts, causing significant damage to agricultural crops in regions where these pests are prevalent. On the other hand, genera like Gregarina and Plasmodium can infect a wide range of insect hosts, including mosquitoes, beetles, and moths.

   The host range of a protozoan pathogen is determined by factors such as host susceptibility, physiological compatibility, and ecological interactions within the insect community. Certain protozoa have evolved mechanisms to exploit specific hosts, while others exhibit opportunistic behavior and can infect a variety of insect species.

   Understanding the host range of insect pathogenic protozoa is essential for predicting disease dynamics, identifying potential reservoir hosts, and developing targeted control strategies to mitigate pest outbreaks in agricultural and natural ecosystems.

b. **Environmental Conditions:** The epizootiology of protozoan-induced diseases is strongly influenced by environmental factors such as temperature, humidity, and host density. Optimal conditions for disease transmission and development vary among different protozoan species, reflecting their adaptation to specific ecological niches.

   For example, epizootics caused by Nosema locustae, a common pathogen of grasshoppers and locusts, are often associated with high temperatures and humidity. These environmental conditions favor spore germination and transmission among grasshopper populations, leading to rapid disease spread and significant reductions in pest numbers.

   In contrast, other protozoan pathogens may thrive in cooler or drier environments, exhibiting seasonal patterns of disease prevalence that coincide with changes in weather conditions. Understanding the environmental requirements of protozoan pathogens is essential for predicting disease outbreaks and implementing timely control measures to prevent economic losses in agriculture and forestry.

c. **Transmission Routes:** Insect pathogenic protozoa can be transmitted through various routes, including ingestion, direct contact, or through intermediate vectors such as parasitic mites or contaminated food sources. The mode of transmission depends on the specific life cycle and biology of the protozoan pathogen, as well as the behavior and ecology of the host insect.

Ingestion of contaminated material, such as plant foliage or soil, is a common mode of transmission for many protozoan pathogens. Protozoan spores or cysts present on the surface of plants or in the soil can be ingested by susceptible insects during feeding, leading to infection of the gut epithelium and subsequent disease development.

Direct contact transmission may occur through physical contact between infected and susceptible insects, particularly in crowded or densely populated insect populations. Infected individuals may shed protozoan spores or cysts into the environment, where they can come into contact with healthy individuals and initiate new infections.

Vector-mediated transmission involves the transfer of protozoan spores or cysts by parasitic organisms, such as mites or other arthropods. These vectors serve as intermediaries in the transmission of protozoan pathogens between infected and susceptible hosts, facilitating the spread of disease within insect populations.

d. **Seasonal Patterns:** The prevalence and intensity of protozoan-induced diseases often exhibit seasonal patterns, with disease outbreaks typically occurring during periods of favorable environmental conditions for protozoan development and transmission. In temperate regions, disease incidence may peak during warm and humid months, coinciding with the peak activity of insect hosts and the proliferation of protozoan pathogens.

   In tropical regions, where environmental conditions are more stable throughout the year, disease transmission may be more constant, with protozoan pathogens persisting in the environment and infecting susceptible hosts on an ongoing basis. However, seasonal fluctuations in rainfall, temperature, and host abundance can still influence the dynamics of disease transmission and the severity of outbreaks.

   Understanding seasonal patterns of disease occurrence is essential for predicting disease risk, implementing preventive measures, and optimizing control strategies to mitigate the impact of protozoan-induced diseases on agricultural production and natural ecosystems.

2. Symptomatology of Diseases Caused by Insect Pathogenic Protozoa

The symptomatology of diseases caused by insect pathogenic protozoa encompasses a wide range of clinical manifestations, reflecting the physiological effects of protozoan infection on the insect host. These symptoms is essential for early detection, diagnosis, and management of protozoan-induced diseases in agricultural and natural ecosystems.

a. Behavioral Changes

Infected insects often exhibit altered behaviors as a result of protozoan infection. These behavioral changes may include:

1. **Lethargy:** Infected insects may become lethargic or sluggish, displaying reduced activity levels compared to healthy individuals. This decrease in activity may be a consequence of the physiological effects of protozoan infection on the insect's metabolism and energy balance.
2. **Reduced Feeding Activity:** Protozoan-infected insects may show reduced appetite or feeding activity. This decrease in feeding behavior may be due to disruptions in the insect's digestive processes, leading to decreased nutrient absorption and energy acquisition.
3. **Abnormal Locomotion:** Infected insects may exhibit abnormal locomotion patterns, such as uncoordinated movements or impaired mobility. These locomotor abnormalities may result from damage to the insect's nervous system or musculature caused by protozoan infection.

Behavioral changes in infected insects can impact their interactions with the environment, including feeding behavior, reproduction, and dispersal, ultimately affecting population dynamics and ecosystem functioning.

b. Physiological Effects: Protozoan-induced diseases can cause a variety of physiological disturbances in infected insects, affecting various organ systems and metabolic processes. These physiological effects may include:

1. **Disruption of Digestive Processes:** Protozoan infections can interfere with the insect's digestive processes, leading to impaired nutrient absorption and utilization. Damage to the gut epithelium or disruption of enzyme activity may result in inefficient digestion and nutrient uptake, compromising the insect's overall health and fitness.
2. **Impaired Growth and Development:** Infected insects may experience delays or abnormalities in growth and development. Protozoan pathogens can disrupt hormonal signaling pathways or interfere with cellular metabolism, leading to developmental defects, stunted growth, or delayed metamorphosis in insect larvae or nymphs.
3. **Damage to Internal Organs:** Severe protozoan infections can cause damage to internal organs, including the gut, fat body, and reproductive organs. Protozoan pathogens may proliferate within host tissues, causing cellular damage, inflammation, or necrosis. Damage

to vital organs can impair physiological functions such as digestion, metabolism, and reproduction, ultimately leading to morbidity and mortality in infected insects.

4. **Systemic Effects:** In severe cases, protozoan infections can have systemic effects on the insect host, affecting multiple organ systems and physiological processes. Systemic infections may result in lethargy, weakness, or paralysis, as well as secondary complications such as secondary infections or metabolic imbalances. Systemic effects of protozoan infection may ultimately lead to death in infected insects.

c. **External Signs** Protozoan-infected insects may exhibit external signs or physical abnormalities that are visible to the naked eye. These external signs may include:

1. **Changes in Coloration:** Infected insects may display alterations in body coloration or pigmentation. Changes in coloration may be indicative of underlying physiological disturbances, such as immune responses or metabolic changes associated with protozoan infection.
2. **Body Size or Morphological Changes:** Protozoan-infected insects may show changes in body size or morphology compared to healthy individuals. These changes may include alterations in body shape, size, or proportions, as well as the development of abnormal appendages or structures.
3. **Visible Lesions or Deformities:** In some cases, visible lesions or deformities may develop on the external surface of infected insects. These lesions may result from tissue damage caused by protozoan proliferation or inflammatory responses mounted by the host immune system. Lesions or deformities may vary in size, shape, and location depending on the site and severity of infection.

External signs of protozoan infection can serve as diagnostic indicators for identifying infected individuals and monitoring disease prevalence within insect populations. Visual inspection of insect specimens can provide valuable information about the severity and progression of protozoan-induced diseases in the field.

d. **Mortality Events:** Severe protozoan infections can lead to mortality events within insect populations, particularly during disease outbreaks or epizootics. Mass die-offs of infected insects may occur, resulting in significant reductions in pest populations and subsequent ecological impacts. Mortality events may be precipitated by factors such as:

1. **Disease Outbreaks:** Protozoan-induced diseases may spread rapidly within insect populations, leading to high mortality rates among infected individuals. Disease outbreaks can occur when environmental conditions are favorable for protozoan transmission and development, resulting in widespread infection and mortality.
2. **Epizootics:** Epizootics are characterized by sudden, large-scale outbreaks of disease within insect populations. Protozoan pathogens can trigger epizootics when host density is high, facilitating rapid disease transmission and amplification within susceptible populations. Epizootics may result in mass mortality events, causing significant disruptions to ecosystem dynamics and agricultural production.
3. **Population Declines:** Prolonged or severe protozoan infections can cause population declines in insect pest species, reducing their abundance and impact on agricultural crops or natural ecosystems. Population declines may have cascading effects on ecosystem processes and community structure, altering predator-prey interactions and ecosystem services provided by insects.

3. Etiology of Diseases Caused by Insect Pathogenic Protozoa

The etiology of diseases caused by insect pathogenic protozoa is essential for elucidating the mechanisms of disease transmission, and host-pathogen interactions, and developing effective control strategies. The etiology encompasses the identification and characterization of the causative agents, the investigation of their life cycles, and the study of host-pathogen interactions.

a. **Pathogen Identification:** The etiology of diseases caused by insect pathogenic protozoa begins with the identification and characterization of the causative agents. Various techniques are employed for pathogen identification, including:

   1. **Molecular Techniques:** Molecular methods such as polymerase chain reaction (PCR), DNA sequencing, and phylogenetic analysis are commonly used to identify protozoan pathogens based on their genetic material. Specific regions of the protozoan genome can be targeted for amplification and sequencing, allowing for accurate species identification and phylogenetic analysis.
   2. **Microscopy:** Microscopic examination of infected insect tissues or environmental samples can provide valuable information about the morphology, size, and characteristics of protozoan pathogens. Staining techniques, such as Giemsa or DAPI staining, can enhance

the visualization of protozoan structures, aiding in their identification and characterization.

3. **Culture-Based Methods:** Culture-based methods involve the isolation and cultivation of protozoan pathogens from infected insect tissues or environmental samples. Cultured isolates can be characterized morphologically, biochemically, and genetically to identify the species and assess their pathogenicity.

Pathogen identification provides the foundation for further studies on disease epidemiology, transmission dynamics, and host-pathogen interactions.

b. **Pathogen Life Cycle:** Understanding the life cycle of protozoan pathogens is essential for elucidating their mode of transmission, host specificity, and pathogenicity. Protozoan life cycles typically involve several stages, including:

1. **Spore Formation:** Many insect pathogenic protozoa produce spores as a dormant, resistant stage of their life cycle. Spores are specialized structures capable of surviving adverse environmental conditions and facilitating transmission between hosts.
2. **Host Invasion:** Protozoan pathogens infect their insect hosts through various routes, such as ingestion, direct contact, or vector-mediated transmission. Once inside the host, protozoa invade host tissues and cells, where they establish infection and replicate.
3. **Intracellular Replication:** Within the host, protozoan pathogens undergo intracellular replication, exploiting host cellular machinery to proliferate and spread. Protozoa may manipulate host cell processes, evade immune defenses, and modulate host physiology to promote their survival and replication.
4. **Transmission:** Protozoan pathogens are transmitted between hosts through various mechanisms, including ingestion of contaminated material, direct contact between infected and susceptible individuals, or vector-mediated transmission by arthropod vectors. Transmission routes influence the epidemiology and spread of protozoan-induced diseases within insect populations.

Each stage of the protozoan life cycle represents potential targets for disease control strategies, including the development of vaccines, antiparasitic drugs, and environmental management practices to disrupt transmission.

c. **Host-Pathogen Interactions:** The etiology of protozoan-induced diseases also encompasses the study of host-pathogen interactions, including host immune responses, pathogen virulence factors, and mechanisms of host resistance or susceptibility. These interactions shape the dynamics of disease transmission and the outcome of infection within insect populations.

1. **Host Immune Responses:** Insect hosts mount immune responses against protozoan pathogens, including cellular and humoral immune defenses. Immune mechanisms such as phagocytosis, encapsulation, and production of antimicrobial peptides play crucial roles in limiting protozoan infection and promoting host survival.
2. **Pathogen Virulence Factors:** Protozoan pathogens produce various virulence factors, including toxins, adhesion molecules, and immune evasion strategies, to facilitate host invasion, colonization, and replication. Virulence factors contribute to the pathogenicity of protozoan infections and influence disease severity and outcome.
3. **Host Resistance and Susceptibility:** Host resistance to protozoan infection is influenced by genetic, physiological, and environmental factors that determine the host's ability to resist or tolerate infection. Conversely, host susceptibility factors may increase the likelihood of protozoan colonization and disease development. Understanding the mechanisms underlying host resistance and susceptibility can inform breeding programs and pest management strategies aimed at enhancing host resistance to protozoan pathogens.

Biological Control of Weeds Using Insects

Weed control is economically and environmentally vital, with herbicides accounting for nearly half of agricultural sales globally. In developed countries, herbicides are the primary method for weed control, although mechanical weeding is becoming more prevalent. In developing countries, hand weeding can constitute up to 60% of preharvest labor. Organic farming in developed countries also relies heavily on hand weeding. If left unchecked, weeds can cause total crop yield loss and severe environmental damage, especially from invasive species. Biological control of weeds has proven to be highly effective. Unlike biological control of insect pests, which often involves conservation and augmentative methods, classical biological control is the primary approach for weeds.

Augmentation and Conservation Biological Control of Weeds

Augmentation biological control involves increasing the population of natural enemies, such as fungi, to control weeds. While much research has been done on using fungi, field applications have been limited. However, some successes include the Chondrilla Rust Fungus (*Puccinia chondrillae*) for controlling

Skeleton Weed in Australia and the fungus *Chondrostereum purpureum* for managing American Bird Cherry (*Prunus serotina*) in Europe.

Native insects are occasionally used in a combination of augmentation and conservation biological control, but practical applications are limited. More commonly, augmentation is used with exotic biological control agents, especially when the agent's natural dispersal is inadequate, and the weed is scattered across discrete areas. Examples include the control of cacti in Australia and South Africa through the redistribution of mealybugs and the management of the floating fern Salvinia molesta by the salvinia weevil (*Cyrtophagous salviniae*).

Classical Biological Control

Classical biological control of weeds, which began in the early 20th century, focuses on introducing natural enemies from the weed's native range. This method has been particularly successful in countries with extensive rangeland, such as the USA, Australia, South Africa, Canada, and New Zealand. For instance, Hawaii has achieved a nearly 50% success rate in controlling 7 out of 21 targeted weed species, with significant partial control of three more. Recently, there has been a shift towards using biological control for weeds in natural ecosystems, known as environmental weeds.

Procedures in Classical Biological Control of Weeds

Classical biological control of weeds is a systematic and strategic approach that involves introducing natural enemies to control invasive weed species. This method is preferred for its potential long-term benefits and minimal environmental impact compared to chemical methods. The process includes several meticulously planned steps:

1. Selection of Weed Targets

The first step in classical biological control is to identify and select weed species as targets based on specific criteria. The selection process is crucial because it determines the project's scope and potential impact. Key factors considered include:

- **Economic and Environmental Impact**: The target weed must be one that causes significant economic loss or environmental damage. The more widespread and damaging the weed, the greater the justification for its control.
- **Likelihood of Success**: Some weeds are more amenable to biological control strategies than others, depending on their ecological characteristics and the availability of natural enemies.

- **Cost-Benefit Analysis**: The expected benefits of controlling the weed are weighed against the costs of developing and implementing a biological control program. This includes both direct costs and potential indirect costs, such as non-target effects.

2. Foreign Exploration

Once a target weed has been selected, the next step is foreign exploration. This phase involves studying the weed in its native habitat to understand its ecology, natural enemies, and conditions promoting its growth. Key activities include:

- **Taxonomic Studies**: Accurate identification of the weed using both classical taxonomy and molecular techniques to ensure the correct species is targeted.
- **Collection of Natural Enemies**: Researchers collect potential biological control agents associated with the weed in its native range. This often includes a variety of insects, mites, and pathogens.
- **Preliminary Screening**: Initial assessments of the collected agents to determine their host specificity and potential impact on the weed.

3. Agent Selection

Choosing the right biological control agent is one of the most critical decisions in the process. This step involves evaluating the effectiveness and safety of potential agents. Considerations include:

- **Host Specificity**: Agents must primarily affect the target weed and not harm non-target species, particularly economically important or ecologically significant plants.
- **Impact on the Weed**: The agent must be capable of significantly reducing the weed's population or its reproductive capacity.
- **Survival and Adaptation**: The agent must be able to survive and establish itself in the target environment outside its native range.

4. Host Specificity Testing

Host specificity testing is a detailed and rigorous phase where potential agents are tested under controlled conditions to confirm that they will not harm non-target plants. This process includes:

- **Laboratory and Greenhouse Tests**: Agents are tested on a range of plant species, especially those closely related to the target weed and economically important crops.

- **Field Tests**: In some cases, field tests are conducted in confined conditions to observe the agents' behavior in a more natural environment but still under strict control.
- **Regulatory Review**: The results of these tests are crucial for obtaining regulatory approval to release the agent.

5. Rearing and Release

Once an agent passes all tests and receives regulatory approval, it is mass-reared in specialized facilities. The rearing process must ensure a high-quality, viable population of the agent. Key considerations include:

- **Scaling Up Production**: Developing methods to produce large numbers of the agent efficiently.
- **Quality Control**: Monitoring health, vigor, and reproductive capacity to ensure the agents are effective upon release.
- **Release Strategies**: Determining optimal times and methods for releasing the agents into the environment to maximize their impact on the weed population.

6. Post-Release Evaluation

After the agents are released, an ongoing evaluation is critical to assess the success of the biological control program. This includes:

- **Monitoring Establishment**: Checking whether the agents establish themselves in the target area and start impacting the weed population.
- **Assessing Impact**: Evaluating the reduction in weed density and spread over time due to the activity of the biological control agents.
- **Long-term Surveillance**: Monitoring the ecosystem for any unforeseen impacts, including potential non-target effects and ecological shifts.

Results Achieved in Weed Biological Control

The efficacy of weed biological control programs is evaluated based on the level of control achieved over the target weed populations. Success is typically categorized into three distinct outcomes: complete control, substantial control, and negligible control. Each category reflects the degree to which the biological control agents affect the target weed and the consequent reduction in the need for other weed management strategies. Here is an in-depth examination of each category:

Complete Control

Complete control is the ideal outcome of a biological control program. In this scenario, the introduced natural enemies effectively suppress the target weed population to the extent that no additional weed control methods are necessary. Achieving complete control means that the biological agents have successfully adapted to the new environment and are efficiently managing the weed population on their own. This level of success often leads to significant economic savings and ecological benefits, including:

- **Reduction in Chemical Use**: The need for herbicides is eliminated, reducing costs and minimizing environmental pollution.
- **Stabilization of Ecosystems**: The controlled weed no longer competes with native species, allowing for ecological restoration and biodiversity enhancement.

Examples of complete control are relatively rare but highly celebrated within the field of biological control. One such example is the control of the floating water fern Salvinia molesta in Australian waterways by the salvinia weevil (Cyrtobagous salviniae), which has led to the recovery of affected aquatic ecosystems.

Substantial Control

Substantial control is achieved when the biological control agents significantly reduce the weed population, but some additional control measures remain necessary. This outcome is more common than complete control and still offers considerable benefits:

- **Reduced Management Costs**: While other control methods are still required, their frequency and intensity are significantly reduced.
- **Improved Control Efficiency**: Biological control works synergistically with other methods, such as mechanical removal or limited herbicide application, to keep weed populations at manageable levels.

Substantial control often occurs in complex ecosystems where multiple factors influence weed dynamics, or when the biological control agent does not entirely eliminate the weed but keeps it below damaging thresholds. An example includes the control of prickly pear cactus in parts of Australia using the cactoblastis moth (Cactoblastis cactorum), which has drastically reduced the cactus population but not entirely eradicated it.

Negligible Control

Negligible control describes situations where the biological control agents have been established but have minimal impact on the overall weed population. In these cases, the primary reliance for weed management continues to be on conventional methods such as chemical herbicides or manual removal. Reasons for negligible control might include:

- **Inadequate Agent Performance**: Sometimes agents do not adapt well to the new environment or fail to thrive in sufficient numbers to impact the weed population.
- **Ecological Mismatches**: The agents might not be perfectly suited for the range of environmental conditions present in the target area, limiting their effectiveness.

Negligible control serves as a learning opportunity for researchers to understand the complexities of biological interactions and refine their strategies for future releases. For instance, certain agents introduced to control invasive thistle species in North America have shown limited success, leading to ongoing research and adaptation of strategies.

Case Studies of Successful Biological Control Programs

The Case of St. John's Wort (*Hypericum perforatum*)

St. John's Wort, a problematic weed in pastures and natural ecosystems, was successfully controlled using Chrysolina beetles (*Chrysolina quadrigemina* and *Chrysolina hyperici*). These beetles feed on the foliage of St. John's Wort, significantly reducing its prevalence.

Prickly Pear Cactus (Opuntia spp.)

Prickly pear cactus invaded large areas of Australia, rendering the land unproductive. The introduction of the *Cactoblastis cactorum* moth from Argentina in the 1920s resulted in dramatic control of the cactus. The larvae burrow into the cactus pads, causing extensive damage and leading to the cactus's decline.

Salvinia molesta

Salvinia molesta, an invasive aquatic fern, became a major problem in waterways. The introduction of the weevil *Cyrtobagous salviniae* in Australia and India successfully controlled this weed. The weevil's larvae feed on the plant's buds and rhizomes, reducing its spread.

Challenges and Limitations of Biological Control of Weeds Using Insects

Biological control of weeds using insects, while offering a sustainable alternative to chemical herbicides, encounters several significant challenges and limitations that can affect its efficacy and acceptance. These challenges encompass ecological, biological, regulatory, and social aspects, each of which requires careful consideration and management to ensure the success and sustainability of biological control programs.

1. Non-Target Effects

One of the principal concerns with using insects for biological control is the potential for non-target effects, where introduced agents negatively affect native or economically important plant species, as well as other organisms within the ecosystem. Ensuring that biological control agents selectively target only the intended weed species is crucial but challenging.

Risks to Non-Target Species:

- **Direct Effects:** These occur when the introduced biological control agent attacks non-target plants or animals. For example, if an insect released to control a particular weed species also feeds on native plants, it can reduce biodiversity and disrupt local ecosystems.
- **Indirect Effects:** These are more subtle and can include changes in ecosystem structure and function, such as alterations in nutrient cycling, hydrology, or the displacement of native species by changing competitive relationships.

Mitigation Strategies

- **Host Specificity Testing:** Rigorous testing in quarantine facilities and controlled field trials helps ensure agents target only specific weed species. This process involves extensive research and often international collaboration.
- **Genetic Studies:** Utilizing genetic and molecular tools to better understand the biology and feeding preferences of potential control agents can refine selections and minimize risks.
- **Regulatory Oversight:** Governments typically require comprehensive risk assessments before approving the release of biological control agents to prevent unintended consequences.

2. Climate and Habitat Compatibility

The success of biological control agents largely depends on their ability to survive and thrive in the local climate and habitat where the target weed is problematic. However, the adaptability of these agents to new environmental conditions is not always guaranteed.

Adaptation Challenges

- **Climatic Factors:** Temperature extremes, precipitation patterns, and seasonal variability can influence the survival and reproductive success of introduced insects. Agents that thrive in their native habitat may struggle in a new climate, leading to poor establishment.

- **Habitat Specificity:** The physical environment, including soil type, water availability, and vegetation structure, must support the lifecycle and behaviors of the control agents. A mismatch can lead to ineffective control efforts.

Adaptation Strategies

- **Ecological Matching:** Selecting agents from regions with climatic and ecological conditions similar to the target area can increase the likelihood of successful adaptation.
- **Acclimatization Trials:** Conducting pilot studies or phased releases can help identify potential adaptation issues before full-scale implementation.
- **Breeding and Genetic Adaptation:** Advanced biotechnological techniques, such as selective breeding or genetic modification, might enhance the adaptability of agents, although these approaches come with their own ethical and regulatory challenges.

3. Resistance Development

Just as pests can develop resistance to chemical pesticides, there is a risk that weeds may evolve resistance to their biological control agents. This can occur through natural selection if there are genetic variations within the weed population that confer resistance to the biological agent.

Resistance Mechanisms

- **Morphological Changes:** Some weeds might evolve physical traits that deter feeding or oviposition by control agents, such as thicker leaves or altered growth forms.
- **Chemical Defenses:** Increased production of secondary metabolites that make the plants less palatable or toxic to the biological agents is another potential resistance mechanism.
- **Reproductive Strategies:** Weeds might also change their reproductive timing or methods to avoid peak periods of biological agent activity.

Management of Resistance

- **Integrated Weed Management:** Combining biological control with other weed management strategies, such as mechanical removal or selective herbicide use, can reduce the selection pressure for resistance.
- **Monitoring and Surveillance:** Regular monitoring of weed response to biological control agents can detect early signs of resistance development, allowing for timely adjustments in management strategies.

- **Diverse Agent Assemblage:** Releasing a combination of different biological control agents that target the weed in various ways can reduce the likelihood of resistance development.

4. Regulatory and Public Acceptance

The introduction of exotic species for biological control involves navigating complex regulatory frameworks designed to safeguard environmental and public health. Additionally, gaining public acceptance is crucial, as stakeholders may have concerns about potential risks associated with releasing non-native organisms into the environment.

Regulatory Challenges

- **Strict Approval Processes:** Obtaining regulatory approval for the release of biological control agents can be a lengthy and expensive process, requiring detailed risk assessments, environmental impact statements, and public consultations.
- **International Coordination:** For weeds that cross national boundaries, coordinated regulatory approaches are necessary, which can complicate the approval processes.

Strategies for Enhancing Acceptance

- **Public Engagement:** Educating the public and stakeholders about the benefits and safety of biological control through workshops, seminars, and transparent communication can alleviate concerns and foster support.
- **Collaboration with Stakeholders:** Involving farmers, conservationists, and other stakeholders in the planning and implementation phases of biological control projects can build trust and improve program designs based on local knowledge and needs.
- **Adaptive Management:** Establishing mechanisms for ongoing assessment and adaptation of biological control programs can assure the public and regulatory bodies that potential problems will be managed responsibly.

Future Directions and Innovations

The future of biological control of weeds using insects looks promising with ongoing research focusing on innovative approaches:

1. **Genomic and Biotechnological Advances**: Understanding the genetic makeup of both weeds and their insect enemies can lead to more targeted and effective control strategies.

2. **Integrated Pest Management (IPM)**: Combining biological control with other weed management strategies, such as cultural practices and minimal herbicide use, to create a holistic approach.
3. **Conservation Biological Control**: Enhancing the effectiveness of native insect populations through habitat manipulation and conservation practices.
4. **Climate Change Adaptation**: Developing biological control programs resilient to the impacts of climate change, ensuring long-term sustainability.

Weed Killers

Insects that help in controlling weeds by feeding on them are called weed killers. These insects play a crucial role in biological weed control, helping manage invasive weed species and reduce their impact on ecosystems and agricultural productivity.

1. *Dactylopius tomentosus* (Cochineal Insect)

Target Weed: Prickly pear (*Opuntia dillenii*)

Introduction and Impact: *Dactylopius tomentosus*, commonly known as the cochineal insect, was introduced into India in 1925 to control the invasive prickly pear cactus. Within 5-10 years, this insect had effectively managed to reduce the population of Opuntia dillenii. The cochineal insect feeds on the sap of the cactus, causing the plant to dehydrate and eventually die. This biological control method proved to be a sustainable and eco-friendly alternative to chemical herbicides.

Identification

- **Appearance:** Small, oval-shaped insects covered in white, cotton-like wax.
- **Lifecycle:** Females lay eggs under the protective wax, which hatch into nymphs that disperse and begin feeding on new plants.

Mechanism of Control: The cochineal insect injects its stylet into the cactus pads and sucks out the sap. This feeding action not only deprives the plant of its vital fluids but also introduces toxic saliva into the plant tissues, leading to necrosis and the eventual death of the cactus. The establishment of *Dactylopius tomentosus* in areas heavily infested with prickly pear has shown significant success, reducing the density of these invasive plants and allowing native vegetation to recover.

2. Aristolochia Butterfly (*Papilio aristolochiae*)

Target Weed: Aristolochia spp.

Introduction and Impact: The Aristolochia butterfly, scientifically known as Papilio aristolochiae, is a natural predator of Aristolochia species, which are considered weeds in many regions. The larvae of this butterfly feed voraciously on the leaves of the Aristolochia plant, reducing its ability to spread and thrive.

Identification

- **Appearance:** Adults are large butterflies with black and white markings. Larvae are brightly colored, often with patterns that warn predators of their toxicity.
- **Lifecycle:** The female butterfly lays eggs on the leaves of the Aristolochia plant, and the emerging larvae feed on the leaves until they pupate.

Mechanism of Control: The Aristolochia butterfly larvae possess strong mandibles that can efficiently consume large amounts of leaf material. This defoliation weakens the plant, reducing its photosynthetic capacity and reproductive potential. Over time, repeated infestations can lead to a significant decline in the weed population, allowing desirable plant species to reestablish themselves in the ecosystem.

3. Calotropis Butterfly (*Danaus chrysippus*)

Target Weed: Calotropis spp.

Introduction and Impact: *Danaus chrysippus*, known as the Calotropis butterfly, feeds on the Calotropis plant, a common weed in various parts of the world. The larvae of this butterfly consume the leaves of Calotropis, which helps in controlling its spread.

Identification

- **Appearance:** Adults are orange with black and white spots. Larvae are striped with black, white, and yellow bands.
- **Lifecycle:** Females lay eggs on Calotropis leaves, and the larvae feed on the leaves and sometimes the flowers.

Mechanism of Control: The Calotropis butterfly larvae feed extensively on the leaves, causing defoliation and reducing the plant's ability to grow and reproduce. The larvae have evolved to tolerate the toxic compounds in Calotropis, using them for their own defense against predators. This selective feeding helps manage Calotropis populations effectively, especially in areas where this weed is problematic.

4. AK Grasshopper (*Poecilocerus pictus*)

Target Weed: Calotropis spp.

Introduction and Impact: The AK grasshopper, *Poecilocerus pictus*, is another effective biological control agent for Calotropis. This grasshopper feeds on the leaves of the Calotropis plant, significantly reducing its growth and spread.

Identification

- **Appearance:** Large, colorful grasshoppers with yellow and green patterns.
- **Lifecycle:** Grasshoppers lay eggs in the soil, and the nymphs feed on Calotropis plants after hatching.

Mechanism of Control: *Poecilocerus pictus* exhibits a preference for Calotropis plants, consuming significant amounts of foliage. The feeding activity of both nymphs and adults weakens the plants, curtailing their growth and reproductive capabilities. By reducing the biomass of Calotropis, these grasshoppers contribute to the control of this weed and support the reestablishment of native vegetation.

5. Water Hyacinth Weevils (*Neochetina eichhorniae* and *Neochetina bruchi*)

Target Weed: Water hyacinth (Eichhornia crassipes)

Introduction and Impact: *Neochetina eichhorniae* and *Neochetina bruchi* are two weevil species used to control the invasive water hyacinth. The larvae of these weevils tunnel into the petioles and feed inside, causing significant damage to the plant. Ten pairs of adults and their progeny can control the growth of water hyacinth in an area of approximately 0.58 m^2.

Identification

- **Appearance:** Small, dark-colored weevils with a distinctive snout.
- **Lifecycle:** Adults lay eggs on the water hyacinth, and the larvae burrow into the plant, feeding on the tissue and causing structural damage.

Mechanism of Control: The larvae of Neochetina spp. feed internally, creating tunnels within the petioles and crown of the water hyacinth. This internal feeding disrupts the vascular system, leading to reduced nutrient and water transport within the plant. Over time, the damaged plants become less buoyant and sink, reducing the coverage of water hyacinth on water bodies. This helps restore aquatic habitats and improves water flow in infested areas.

6. Parthenium Weed Killer (*Zygogramma bicolorata*)

Target Weed: *Parthenium hysterophorus*

Introduction and Impact: Zygogramma bicolorata, a leaf beetle, is used to control the invasive Parthenium weed. Both adults and grubs feed on the leaves and flowers of the Parthenium plant. Two beetles can destroy one plant in approximately 45 days, making them highly effective in reducing the weed population.

Identification

- **Appearance:** Adults are small, brownish beetles with distinct black spots. The grubs are yellowish and soft-bodied.
- **Lifecycle:** Females lay eggs on Parthenium leaves, and the emerging grubs feed on the leaves and flowers, causing extensive damage.

Mechanism of Control: The larvae and adult beetles of *Zygogramma bicolorata* feed on the foliage of *Parthenium hysterophorus*, leading to defoliation and reduced flowering. This feeding activity not only weakens the plants but also disrupts their reproductive cycle, preventing the production of seeds. Over time, the cumulative effect of this biological control agent can lead to a significant decline in Parthenium populations, allowing native and cultivated plants to thrive.

Ecological and Agricultural Benefits of Phytophagous Natural Enemies

The use of phytophagous natural enemies for weed control offers several ecological and agricultural benefits:

1. **Sustainable Weed Management:** Biological control methods provide a long-term solution to weed problems without the need for repeated chemical applications.
2. **Environmental Safety:** Unlike chemical herbicides, natural enemies do not leave harmful residues in the soil, water, or air, reducing the risk of environmental contamination.
3. **Biodiversity Conservation:** By targeting specific weed species, natural enemies help maintain biodiversity by allowing native plants to recover and thrive.
4. **Cost-Effectiveness:** Once established, natural enemies can provide ongoing weed control with minimal maintenance costs compared to chemical herbicides.
5. **Reduction of Herbicide Resistance:** The use of natural enemies helps mitigate the development of herbicide-resistant weed populations, a growing concern in many agricultural systems.

Challenges and Considerations in Biological Weed Control

While biological control of weeds using phytophagous insects has numerous advantages, it also presents several challenges and considerations:

1. **Host Specificity:** Ensuring that introduced natural enemies are specific to the target weed and do not harm non-target plant species is crucial.
2. **Climate Adaptation:** Natural enemies must be able to survive and thrive in the climatic conditions of the area where they are introduced.
3. **Monitoring and Evaluation:** Continuous monitoring is necessary to assess the effectiveness of biological control agents and their impact on the ecosystem.
4. **Integration with Other Methods:** Combining biological control with other weed management practices can enhance overall effectiveness and sustainability.
5. **Public Perception and Regulation:** Gaining public acceptance and navigating regulatory frameworks for the introduction of biological control agents require careful planning and communication.

Defense Mechanisms in Insects Against Pathogens

Insects, as integral components of terrestrial ecosystems, encounter a myriad of plant pathogens in their environment. These pathogens pose significant threats to both wild and cultivated plants, thereby impacting ecosystem dynamics, agricultural productivity, and food security. In response to the constant pressure exerted by plant pathogens, insects have evolved a diverse array of defense mechanisms aimed at mitigating the detrimental effects of pathogen attack. These defense strategies encompass both physical and chemical barriers, as well as complex immune responses that contribute to the overall health and survival of insect populations.

1. Physical Barriers

Insects, being ubiquitous inhabitants of terrestrial ecosystems, have evolved an impressive array of defense mechanisms to combat the constant threat posed by plant pathogens. Among these defenses, physical barriers play a fundamental role in preventing pathogen entry and maintaining the integrity of the insect's body.

a. **Exoskeleton:** The exoskeleton serves as the first line of defense against plant pathogens for insects. Composed primarily of chitin, a sturdy polysaccharide, and various proteins, the exoskeleton forms a rigid outer covering that provides mechanical protection against physical injury and pathogen penetration. This protective layer, known as the cuticle, acts as a formidable barrier to prevent the entry of fungal spores, bacteria, and other pathogens into the insect's body.

Structure of the Cuticle: The cuticle is a complex structure comprising multiple layers, each serving specific functions in defense against pathogens. The outermost layer, known as the epicuticle, is composed of lipids and waxes that confer hydrophobic properties, reducing water loss and preventing pathogen adherence. Beneath the epicuticle lies the exocuticle, which contains chitin fibers embedded in a protein matrix, providing mechanical strength and rigidity to the exoskeleton. The endocuticle, located closest to the insect's body, consists of chitin and proteins arranged in a less organized manner, contributing to flexibility and elasticity.

Barrier Function: The cuticle acts as a formidable barrier against plant pathogens by physically impeding their entry into the insect's body. Fungal spores, bacteria, and other microorganisms must overcome this barrier to establish infection, making it a critical component of insect defense. The structural integrity of the cuticle is essential for maintaining effective protection against pathogens, highlighting the importance of proper cuticle development and maintenance in insect health and survival.

b. **Waxy Secretions:** Many insects secrete waxes and other hydrophobic compounds onto their body surface, forming a protective barrier against water loss and pathogen ingress. These waxy coatings play a crucial role in reducing the adhesion of fungal spores and bacteria to the insect's cuticle, thereby limiting the establishment of infection. Additionally, waxy secretions enhance the insect's ability to repel water, preventing the accumulation of moisture that could promote fungal growth and pathogen proliferation.

Composition and Function of Waxy Secretions: Insect waxes are composed of a complex mixture of lipids, including hydrocarbons, fatty acids, alcohols, and esters, which contribute to their hydrophobic properties. These waxy coatings form a physical barrier that prevents water and pathogens from adhering to the insect's body surface. By reducing surface wetness, waxy secretions inhibit the germination and attachment of fungal spores and bacteria, limiting their ability to colonize the insect's cuticle and penetrate its tissues.

Adaptations for Water Conservation: In addition to their role in pathogen defense, waxy secretions help insects conserve water in arid or semi-arid environments. By forming a waterproof layer on the cuticle, waxes reduce transpiration and minimize water loss through the integument, allowing insects to survive in dry habitats with limited access to moisture. This dual function of waxy secretions highlights their importance in insect adaptation to diverse environmental conditions and their significance in maintaining insect health and fitness.

c. **Behavioral Adaptations:** In addition to structural defenses, some insects exhibit behavioral adaptations to avoid or minimize exposure to plant pathogens. These behavioral strategies enable insects to reduce their risk of infection and maintain their cuticular integrity and hygiene.

Selective Feeding: Certain herbivorous insects have evolved the ability to selectively feed on plant tissues that are less susceptible to pathogen attack. By avoiding heavily infected or diseased plant parts, insects can minimize their exposure to pathogens and reduce the likelihood of infection. This selective feeding behavior not only helps insects avoid potential harm but also contributes to the overall health and survival of insect populations by minimizing the spread of plant diseases.

Grooming Behaviors: Grooming behaviors, such as self-cleaning and removal of fungal spores from the body surface, play a crucial role in maintaining the cuticular integrity and hygiene of insects. Insects use their legs, mouthparts, and specialized grooming structures to remove debris, dirt, and pathogens from their body surface, thereby reducing the risk of infection. Grooming behaviors are particularly important for insects that inhabit environments with high pathogen loads, such as leaf litter, soil, or decaying organic matter, where the risk of pathogen exposure is elevated.

Social Interactions: In social insect colonies, such as ants and termites, collective behaviors contribute to the defense against plant pathogens. Social grooming, hygienic behaviors, and chemical signaling mechanisms help to detect and remove infected individuals or contaminated materials from the colony environment, thereby reducing the spread of plant pathogens among nestmates. By collectively maintaining colony hygiene and health, social insects enhance their resilience to pathogen outbreaks and promote the survival of the colony as a whole.

2. Chemical Defenses

In addition to physical barriers and behavioral adaptations, insects employ a diverse array of chemical defenses to combat plant pathogens. These chemical compounds, produced either by the insects themselves or acquired from their host plants, play a critical role in deterring pathogen growth and colonization.

Among the various chemical defense mechanisms, secondary metabolites and antimicrobial peptides (AMPs) are particularly noteworthy for their antimicrobial properties and their contribution to insect immunity against plant pathogens.

a. **Secondary Metabolites Diversity and Synthesis:** Insects produce a vast array of secondary metabolites, including alkaloids, terpenoids, phenolics, and other bioactive compounds, which exhibit antimicrobial properties against plant pathogens. These secondary metabolites are synthesized in specialized glands or tissues and serve diverse ecological functions, including defense against predators, communication, and host plant utilization. Insects possess enzymatic machinery and metabolic pathways that enable them to synthesize a wide range of secondary metabolites, which can vary significantly between insect species and developmental stages.

Role in Defense Against Plant Pathogens: Secondary metabolites play a crucial role in defense against plant pathogens by deterring pathogen growth, inhibiting colonization, and reducing disease severity. Many of these compounds exhibit broad-spectrum antimicrobial activity against bacteria, fungi, and other pathogens, making them effective defenses against a wide range of plant pathogens. Alkaloids, such as pyrrolizidine alkaloids and quinolizidine alkaloids, have been shown to inhibit fungal growth and spore germination, while terpenoids, including monoterpenes and sesquiterpenes, possess antifungal and antibacterial properties. Phenolic compounds, such as flavonoids and tannins, can disrupt microbial cell membranes and inhibit enzymatic activity, thereby impairing pathogen growth and virulence.

Sequestration from Host Plants: Some insects have evolved the ability to sequester plant-derived secondary metabolites from their host plants and utilize them as chemical defenses against pathogens. By feeding on plants rich in bioactive compounds, such as alkaloids and terpenoids, insects can accumulate these compounds in their tissues and incorporate them into their defensive arsenal against pathogens. Sequestration of plant secondary metabolites provides insects with a potent means of defense against plant pathogens while reducing the metabolic costs associated with de novo synthesis of these compounds.

b. **Antimicrobial Peptides (AMPs) Innate Immune Response:** The innate immune system of insects produces antimicrobial peptides (AMPs) in response to pathogen infection, providing rapid and effective defense against plant pathogens. AMPs are small, cationic peptides that exhibit broad-spectrum antimicrobial activity against bacteria, fungi, and viruses. These peptides are synthesized by various immune cells, including hemocytes and epithelial cells, and are released into the hemolymph or secreted onto epithelial surfaces in response to pathogen challenge.

Mechanism of Action: AMPs exert their antimicrobial effects by disrupting pathogen cell membranes, inhibiting cell wall synthesis, and modulating immune signaling pathways. These peptides possess amphipathic properties, enabling them to interact with microbial membranes and induce pore formation or membrane disruption, leading to cell lysis and death. Additionally, AMPs can inhibit microbial growth by interfering with essential cellular processes, such as protein synthesis or DNA replication, thereby impairing pathogen viability and proliferation.

Enhanced Resistance to Plant Pathogens: The production of AMPs enhances the insect's resistance to plant pathogens by providing rapid and effective defense against microbial invaders. AMPs act synergistically with other components of the insect immune system, such as phagocytosis and melanization, to eliminate pathogens and limit infection. Moreover, AMPs play a crucial role in modulating immune signaling pathways and activating immune responses, thereby enhancing the overall resilience of the insect host to plant pathogens.

3. Immune Responses

Insects possess a sophisticated immune system that enables them to mount robust defense responses against invading plant pathogens. These immune responses encompass both cellular and humoral components, which work in concert to detect, neutralize, and eliminate pathogens from the insect's body.

a. **Cellular Immunity Mediated by Hemocytes:** Insects possess a cellular immune response mediated by specialized immune cells called hemocytes, which circulate in the hemolymph, the insect's equivalent of blood. Hemocytes play a central role in detecting and eliminating invading pathogens through processes such as phagocytosis, encapsulation, and melanization.

Recognition of Pathogens: Upon encountering pathogen-associated molecular patterns (PAMPs) on the surface of invading pathogens, hemocytes recognize these foreign molecules as non-self and initiate an immune response. PAMP recognition triggers signaling pathways that activate hemocytes and stimulate their effector functions.

Phagocytosis: One of the primary functions of hemocytes in cellular immunity is phagocytosis, the process by which hemocytes engulf and internalize pathogens for destruction. Phagocytic hemocytes recognize and engulf pathogens, forming phagosomes that fuse with lysosomes containing digestive enzymes, resulting in the degradation of the pathogen.

Encapsulation: In cases where phagocytosis is not sufficient to eliminate pathogens, hemocytes can form multicellular aggregates around larger pathogens, a process known as encapsulation. Encapsulation involves the adherence of hemocytes to the surface of the pathogen, followed by the deposition of melanin and other immune factors to isolate and immobilize the invader.

Melanization: Melanization is a potent antimicrobial mechanism in insects, whereby melanin is deposited onto the surface of pathogens to enhance their destruction. Melanin production is triggered by activation of the phenoloxidase enzyme cascade, resulting in the conversion of phenolic compounds into melanin at the site of pathogen infection.

Rapid and Effective Defense: Cellular immunity provides insects with a rapid and effective defense against plant pathogens by directly targeting and eliminating invading microbes. Hemocytes serve as the frontline defenders of the insect's immune system, preventing the proliferation and dissemination of pathogens within the insect's body.

b. **Humoral Immunity Production of Antimicrobial Factors:** Humoral immunity in insects involves the production and secretion of soluble factors, including antimicrobial peptides (AMPs), reactive oxygen species (ROS), and other immune effectors, into the hemolymph. These antimicrobial factors act synergistically to neutralize pathogens, inhibit their growth, and promote their clearance from the insect's body.

AMPs: Antimicrobial peptides (AMPs) are small, cationic peptides with broad-spectrum antimicrobial activity against bacteria, fungi, and viruses. Produced by various immune tissues, including the fat body and hemocytes, AMPs disrupt pathogen cell membranes, inhibit cell wall synthesis, and modulate immune signaling pathways to enhance the insect's resistance to infection.

ROS: Reactive oxygen species (ROS), such as superoxide radicals and hydrogen peroxide, are generated by immune cells and serve as potent antimicrobial agents in insects. ROS production is induced in response to pathogen challenge and contributes to the oxidative damage and eventual destruction of invading pathogens.

Regulation by Signaling Pathways: The humoral immune response in insects is regulated by various signaling pathways, including the Toll, Imd (Immune Deficiency), and JAK/STAT (Janus Kinase/Signal Transducer and Activator of

Transcription) pathways. These signaling pathways coordinate the expression of immune genes in response to pathogen challenge, orchestrating the synthesis and secretion of antimicrobial factors to combat infection.

Synergistic Defense Mechanisms: Humoral immunity acts synergistically with cellular immunity to provide comprehensive defense against plant pathogens. While cellular immunity targets and eliminates pathogens directly, humoral immunity enhances the insect's overall resilience to infection by neutralizing pathogens, inhibiting their growth, and promoting their clearance from the insect's body.

4. Symbiotic Associations

Insects often form symbiotic relationships with microorganisms, including bacteria and fungi, which play crucial roles in protecting the insect host against plant pathogens. These symbiotic associations contribute to the insect's defense mechanisms by producing antimicrobial compounds, competing for resources with pathogenic organisms, and modulating host immune responses.

a. **Endosymbiotic Bacteria:** Contribution to Host Defense: Many insects harbor endosymbiotic bacteria within their bodies, which contribute to host defense against plant pathogens. These symbiotic bacteria reside intracellularly or extracellularly in specialized host tissues and organs and interact closely with the insect host to confer protection against microbial invaders.

Production of Antimicrobial Compounds: Endosymbiotic bacteria produce a variety of antimicrobial compounds, including antibiotics and bacteriocins, which inhibit the growth and proliferation of plant pathogens. These antimicrobial substances help to maintain the balance of microbial communities within the insect host and prevent the establishment of pathogenic infections.

Competition for Resources: Endosymbiotic bacteria compete with pathogenic organisms for resources within the insect's body, including nutrients and ecological niches. By outcompeting plant pathogens for essential resources, symbiotic bacteria limit their growth and colonization, thereby reducing the risk of infection and disease development in the insect host.

Modulation of Host Immune Responses: Endosymbiotic bacteria modulate host immune responses to enhance resistance to infection by plant pathogens. These symbiotic microbes interact with the insect's immune system to stimulate the production of antimicrobial peptides, reactive oxygen species, and other immune effectors, thereby bolstering the insect's defense mechanisms against microbial invaders.

Example: Aphid Endosymbionts: A classic example of endosymbiotic bacteria providing protection against plant pathogens is found in aphids. Aphids harbor endosymbiotic bacteria, such as Buchnera aphidicola, which produce antibiotics that protect the insect host from fungal pathogens transmitted by plants. These symbiotic bacteria play a crucial role in maintaining the health and survival of aphids in diverse ecological environments.

b. **Mycetocytes and Bacteriocytes Specialized Symbiotic Cells:** Certain insects possess specialized cells called mycetocytes and bacteriocytes, which house endosymbiotic fungi or bacteria, respectively. These symbiotic microorganisms reside within the specialized host cells and interact closely with the insect host to provide protection against plant pathogens.

Antimicrobial Activity: Mycetocyte-associated fungi and bacteriocyte-associated bacteria produce antimicrobial compounds that inhibit the growth and proliferation of plant pathogens in the insect host. These symbiotic microorganisms synthesize a diverse array of secondary metabolites, including antibiotics and antifungal compounds, which contribute to the insect's defense mechanisms against microbial invaders.

Detoxification of Allelochemicals: Mycetocytes and bacteriocytes play a crucial role in detoxifying plant allelochemicals ingested by the insect host, which may have antimicrobial properties or be toxic to the insect's physiology. Symbiotic microorganisms metabolize and neutralize these allelochemicals, thereby protecting the insect host from their deleterious effects and enhancing its resistance to plant pathogens.

Modulation of Immune Responses: Mycetocytes and bacteriocytes modulate host immune responses to promote resistance to infection by plant pathogens. These specialized symbiotic cells interact with the insect's immune system to stimulate the production of antimicrobial peptides, detoxifying enzymes, and other immune effectors, thereby enhancing the insect's ability to combat microbial invaders.

Example: Mycetocyte-Associated Fungi: Insects such as beetles and termites possess mycetocytes that house endosymbiotic fungi, such as Pseudogymnoascus, which produce antifungal metabolites that inhibit the growth of plant pathogens in the insect's gut. These symbiotic fungi play a crucial role in protecting the insect host against fungal infections and maintaining its health and survival in diverse ecological habitats.

5. Behavioral Immunity

Behavioral Immunity in Insects Against Plant Pathogens

Behavioral immunity encompasses a range of adaptive behaviors exhibited by insects to enhance their resistance to plant pathogens. These behaviors, including self-medication and social immunity, play a crucial role in protecting individual insects and entire colonies against microbial invaders.

a. Self-Medication

Self-Recognition of Pathogen Threat: Some insects exhibit behaviors indicative of self-medication, where they actively seek out and consume plant secondary metabolites with antimicrobial properties to combat pathogen infection. This phenomenon, known as "zoopharmacognosy," involves the ability of insects to recognize and respond to the presence of pathogens in their environment.

Selective Acquisition of Chemical Defenses: Insects selectively acquire chemical defenses from their host plants, targeting specific secondary metabolites known to possess antimicrobial properties. By consuming plant compounds such as alkaloids, terpenoids, and phenolics, insects bolster their resistance to plant pathogens and mitigate the risk of infection.

Example: Monarch Butterflies: An illustrative example of self-medication in insects is observed in monarch butterflies. Monarchs consume milkweed plants containing cardenolides, a group of secondary metabolites with potent antimicrobial properties. These cardenolides protect monarchs against protozoan parasites, such as Ophryocystis elektroscirrha, thereby reducing the incidence of parasitic infections and enhancing the fitness of the butterfly population.

b. **Social Immunity Collective Defense Mechanisms:** In social insect colonies, such as ants and bees, collective behaviors contribute to the defense against plant pathogens. Social immunity encompasses a suite of behaviors and physiological adaptations that operate at the colony level to detect, neutralize, and eliminate pathogens from the colony environment.

Social Grooming and Hygienic Behaviors: Social insects engage in grooming behaviors to remove pathogens and contaminants from their bodies and nestmates. Social grooming helps to maintain colony hygiene and reduce the transmission of plant pathogens within the colony. Additionally, hygienic behaviors, such as removing diseased or dead individuals from the nest, further limit the spread of pathogens and enhance colony health.

Chemical Signaling and Antimicrobial Compounds: Chemical signaling mechanisms play a crucial role in coordinating social immunity within insect colonies. Pheromones and other chemical cues enable colony members to detect and respond to pathogen threats, facilitating collective defense strategies. Furthermore, symbiotic microbes associated with social insects produce antimicrobial compounds that contribute to colony-level immunity against plant pathogens.

Example: Ant Colonies: Ant colonies exemplify the effectiveness of social immunity in protecting against plant pathogens. Ants engage in grooming behaviors to remove fungal spores and other contaminants from their bodies and nestmates, thereby reducing the risk of disease transmission within the colony. Additionally, ants produce antimicrobial peptides and form mutualistic associations with bacteria that produce antimicrobial compounds, further enhancing colony resilience to plant pathogens.

Multiple Choice Questions (MCQs)

1. **Which behavior is commonly observed in predatory insects when searching for prey?**

   A) Hibernation

   B) Host-seeking

   C) Mating dances

   D) Sun basking

   Answer: B) Host-seeking

2. **What is the primary mode of action of insect pathogenic bacteria?**

   A) They compete for nutrients with the host.

   B) They produce toxins that paralyze the host.

   C) They induce hyperparasitism.

   D) They produce enzymes that dissolve the host's tissues.

   Answer: B) They produce toxins that paralyze the host.

3. **Which insect group is primarily used in the biological control of weeds?**

   A) Bees

   B) Butterflies

   C) Beetles

   D) Moths

   Answer: C) Beetles

4. **What is 'epizootiology'?**
   A) Study of epidemics among plants
   B) Study of animal behavior
   C) Study of disease outbreaks among animal populations
   D) Study of symbiotic relationships in ecosystems

   Answer: C) Study of disease outbreaks among animal populations

5. **Which of the following is not a defense mechanism in insects against pathogens?**
   A) Grooming
   B) Producing antimicrobial peptides
   C) Emitting light signals
   D) Encapsulation of pathogens

   Answer: C) Emitting light signals

6. **What role do nematodes play in controlling insect pests?**
   A) Predation
   B) Competing for resources
   C) Pollination
   D) Decomposition

   Answer: A) Predation

7. **How do insect pathogenic fungi infect their host?**
   A) By altering the host's behavior to make it more susceptible to predators
   B) By penetrating the host cuticle with hyphae
   C) By outcompeting the host for nutrients
   D) By producing bright colors that attract predators

   Answer: B) By penetrating the host cuticle with hyphae

8. **Which of the following pathogens does not typically infect insects?**
   A) Viruses
   B) Protozoa
   C) Fungi
   D) Helminths

   Answer: D) Helminths

9. **What is symptomatology in the context of insect diseases?**

A) The study of symptoms that indicates a specific disease

B) The method of treating insect diseases

C) The science of creating symptoms artificially

D) The evolutionary study of symptoms

Answer: A) The study of symptoms that indicates a specific disease

10. **How do protozoa typically infect their insect hosts?**

A) By attaching to the exterior body and absorbing nutrients

B) Through oral ingestion and then multiplying inside the host

C) By injecting toxic compounds directly into the host

D) By altering the host's DNA

Answer: B) Through oral ingestion and then multiplying inside the host

11. **Which is an example of a behavioral adaptation in parasitic insects?**

A) Feeding at night to avoid predators

B) Developing resistance to pesticides

C) Migrating to cooler areas during summer

D) Changing color to match the environment

Answer: A) Feeding at night to avoid predators

12. **What factor does not influence the epizootiology of insect diseases?**

A) Climate change

B) Host density

C) Topography of the area

D) Color of the host insects

Answer: D) Color of the host insects

13. **Which group of organisms is used in the biological control of mosquito larvae?**

A) Fungi

B) Protozoa

C) Nematodes

D) Viruses

Answer: C) Nematodes

14. **What is the primary effect of insect pathogenic viruses on their host?**

A) Causing sterility in the host

B) Making the host more aggressive

C) Killing the host rapidly

D) Inducing behavioral changes to increase sunlight exposure

Answer: C) Killing the host rapidly

15. **Which strategy is least likely to be used by a parasitic insect?**

A) Laying eggs inside or on the surface of a host

B) Feeding on plant nectar

C) Injecting venom to paralyze the host

D) Mimicking the appearance of the host

Answer: B) Feeding on plant nectar

16. **How do entomopathogenic fungi disperse to new hosts?**

A) By swimming through water

B) By wind dispersal of spores

C) By burrowing through the soil

D) By making loud noises to attract new hosts

Answer: B) By wind dispersal of spores

17. **What is the role of symptomatology in managing insect-borne diseases?**

A) Identifying the correct chemical treatment

B) Predicting the spread of disease

C) Monitoring the outbreak of diseases

D) All of the above

Answer: D) All of the above

18. **Which method is not a typical mode of action for insect pathogenic protozoa?**

A) Parasitizing the host's gut cells

B) Producing light to attract more hosts

C) Causing malnutrition by consuming the host's food

D) Disrupting the host's reproductive system

Answer: B) Producing light to attract more hosts

19. In biological control, what is the primary role of using specific insects against weeds?

A) To fertilize the soil

B) To physically remove the weeds

C) To feed on the weeds and reduce their viability

D) To introduce diseases to the weeds

Answer: C) To feed on the weeds and reduce their viability

20. Which is not a factor controlling the spread of diseases in insect populations?

A) Genetic diversity of the host population

B) Availability of suitable habitats

C) Presence of natural predators

D) Lunar cycles

Answer: D) Lunar cycles

21. What defense mechanism involves insects coating themselves with substances to avoid pathogen entry?

A) Camouflage

B) Grooming

C) Encapsulation

D) Biochemical secretion

Answer: D) Biochemical secretion

22. Which insect behavior is an adaptation to avoid pathogen infection?

A) Aggregation in large numbers

B) Solitary living during high disease prevalence

C) Increased mating activities

D) Decreased movement during daytime

Answer: B) Solitary living during high disease prevalence

23. Which insect group is not typically involved in the biological control of agricultural pests?

A) Ladybugs

B) Dragonflies

C) Wasps

D) Crickets

Answer: D) Crickets

24. What is a common symptom of fungal infection in insects?

A) Increased body temperature

B) Formation of fungal outgrowths

C) Enhanced color vision

D) Increased body size

Answer: B) Formation of fungal outgrowths

25. What factor most influences the effectiveness of biological control agents?

A) The color of the agents

B) The genetic diversity of the agents

C) The age of the agents

D) The time of day they are released

Answer: B) The genetic diversity of the agents

Short Answer Questions

1. Describe the host-seeking behavior of predatory insects.
2. What is the primary mode of action of insect pathogenic nematodes against pests?
3. List three types of pathogens that commonly infect insects and briefly describe one.
4. How do insects utilize physical barriers as a defense mechanism against pathogens?
5. What is epizootiology and why is it important in the study of insect populations?

Long Answer Questions

1. Explain the various strategies parasitic insects use to locate and exploit their hosts. Include examples of specific adaptations that enhance their effectiveness.
2. Discuss the role of insect pathogenic fungi in biological control, focusing on their modes of action and factors that affect their efficacy in different environmental conditions.
3. Describe the principles and methods used in the biological control of weeds using insects. Discuss the ecological considerations and potential impacts on non-target species.

Unit III

Mass production of quality bio-control agents- techniques, formulations, economics, field release/ application and evaluation. Development of insectaries, and their maintenance.

A. Mass Production of Quality Bio-control Agents: Techniques, Formulations, Economics, Field Release/Application, and Evaluation

Biological control agents, such as phytophagous insects, play a significant role in integrated pest management by targeting and suppressing weed populations naturally. Effective utilization of these agents requires understanding their mass production, formulation, economics, field release/application, and evaluation.

Techniques for Mass Production of Bio-control Agents

Mass production of bio-control agents involves rearing insects or other organisms in large quantities while maintaining their effectiveness and quality. The following are key techniques used in mass production:

1. **Laboratory Rearing:** Laboratory rearing of bio-control agents is a meticulous process that ensures the continuous production of high-quality insects under controlled environmental conditions. This technique involves creating optimal conditions for temperature, humidity, and light to simulate natural habitats. Nutritional requirements are met by providing specific diets, which can be natural or artificial, tailored to the needs of the bio-control agents. Breeding protocols are established to manage mating, egg-laying, and larval development cycles efficiently. By controlling these factors, laboratory rearing ensures a stable and sustainable supply of bio-control agents for field application.
2. **Field Collection and Augmentation:** Field collection involves identifying natural populations of bio-control agents in their native habitats. These agents are then collected using various methods such as traps, sweep nets, or hand-picking. Once collected, these agents are used to augment field populations by periodically releasing them in targeted areas. This technique not only boosts the existing populations but also helps in establishing a self-sustaining population of bio-control agents in

the field. Augmentation is particularly useful in areas where the natural population of bio-control agents is insufficient to manage the weed problem effectively.

3. **Insectaries:** Insectaries are dedicated facilities designed specifically for the mass production of bio-control agents. These facilities are equipped with isolation units to prevent contamination and cross-breeding with non-target species. Insectaries maintain optimal environmental conditions and provide high-quality diets to ensure the health and productivity of bio-control agents. The use of insectaries allows for the large-scale production of bio-control agents in a controlled and consistent manner, making it possible to meet the demands of large-scale biological control programs.

Formulations of Bio-control Agents

Formulations enhance the stability, efficacy, and ease of application of bio-control agents. Common formulations include:

1. **Liquid Suspensions:** Liquid suspensions involve suspending bio-control agents in a liquid medium. This formulation is advantageous for its ease of application using standard spraying equipment. Liquid suspensions can be evenly distributed over large areas, ensuring that the bio-control agents are applied directly to the target weeds. This method is particularly effective for agents that need to be applied to foliage or water surfaces.
2. **Granular Formulations:** Granular formulations involve mixing bio-control agents with granular carriers such as clay or sand. This type of formulation is suitable for soil application and provides controlled release of the agents over time. Granular formulations are particularly useful for targeting soil-dwelling weed seeds or roots, ensuring that the bio-control agents remain active in the soil environment for extended periods.
3. **Encapsulation:** Encapsulation involves enclosing bio-control agents in a protective coating, such as alginate or gelatin. This formulation protects the agents from environmental stress and improves their shelf-life. Encapsulation also allows for the controlled release of the agents, ensuring that they remain effective for a longer duration. This method is particularly beneficial for bio-control agents that are sensitive to environmental conditions.
4. **Freeze-dried Powders:** Freeze-drying involves dehydrating bio-control agents and converting them into a powder form. This formulation has a

long shelf-life and is easy to store and transport. Freeze-dried powders can be rehydrated before application, ensuring that the bio-control agents are active and effective. This formulation is particularly useful for distributing bio-control agents to remote or hard-to-reach areas.

Economics of Bio-control Agents

The economic viability of bio-control programs involves several factors:

1. **Production Costs:** Production costs include the initial investment required for setting up insectaries or rearing facilities, as well as ongoing operational expenses. These expenses cover maintaining optimal environmental conditions, providing high-quality diets, and employing skilled labor to manage the production process. While the initial setup costs can be significant, the long-term benefits of sustainable weed control often outweigh these expenses.
2. **Cost-Benefit Analysis:** A cost-benefit analysis compares the effectiveness and long-term benefits of bio-control agents to traditional chemical herbicides. While chemical herbicides may provide quick results, they often require repeated applications and can have adverse environmental effects. In contrast, bio-control agents offer a sustainable and environmentally friendly solution, reducing the need for repeated interventions and minimizing ecological impacts.
3. **Market Pricing:** Setting competitive prices for bio-control agents is essential to ensure their adoption by farmers and land managers. Market pricing should consider production costs, demand, and the comparative cost of chemical herbicides. Additionally, government subsidies and incentives can promote the use of biological control methods, making them more economically attractive.

Field Release/Application of Bio-control Agents

Effective field release and application strategies ensure the successful establishment and impact of bio-control agents:

1. **Release Techniques:** Different release techniques are employed based on the target weed and the specific bio-control agent. Direct release involves dispersing agents directly onto the weed population. Inoculative release introduces a small number of agents to establish a self-sustaining population, while inundative release involves releasing large numbers of agents to achieve immediate control. Each technique has its advantages and is chosen based on the specific requirements of the control program.

2. **Timing and Frequency:** The timing and frequency of releases are critical to the success of bio-control programs. Releases should be timed to coincide with the life cycle of the target weed, ensuring that the bio-control agents are most effective. The frequency of releases depends on the population dynamics of both the weed and the bio-control agent. Regular monitoring and evaluation help determine the optimal timing and frequency for releases.
3. **Application Equipment:** The choice of application equipment depends on the formulation of the bio-control agents. Hand-held or mechanized sprayers are used for liquid formulations, while specialized dispersal devices are employed for granular or encapsulated formulations. Proper calibration and maintenance of equipment ensure that bio-control agents are applied effectively and efficiently.

Evaluation of Bio-control Programs

Continuous evaluation is crucial to measure the success and impact of bio-control programs:

1. **Monitoring Techniques:** Regular field surveys and sampling methods such as quadrats, transects, or sweep nets are used to collect data on weed density and bio-control agent populations. Monitoring helps track the establishment and performance of bio-control agents, providing valuable information for adjusting management strategies.
2. **Effectiveness Metrics:** Effectiveness metrics include quantifying reductions in weed biomass and coverage, as well as evaluating the survival, reproduction, and dispersal of bio-control agents. These metrics provide a comprehensive assessment of the success of bio-control programs and help identify areas for improvement.
3. **Environmental Impact:** Assessing the environmental impact of bio-control programs involves monitoring non-target effects and changes in biodiversity. Ensuring that bio-control agents do not negatively impact non-target species or ecosystems is critical for the long-term success and acceptance of biological control methods.

Case Studies of Bio-control Agents

1. *Dactylopius tomentosus* for Prickly Pear Control

Background: *Dactylopius tomentosus* was introduced to India in 1925 to control prickly pear (*Opuntia dillenii*). The cochineal insect feeds on the sap of the cactus, causing it to dehydrate and die.

Mass Production Techniques

- **Laboratory Rearing:** Maintaining colonies under controlled conditions.
- **Field Augmentation:** Periodic release of lab-reared individuals to boost field populations.

Formulation

- **Liquid Suspension:** For easy application on cactus pads.

Economics

- **Cost-Benefit Analysis:** Significant reduction in prickly pear populations with minimal environmental impact.

Field Release/Application

- **Direct Release:** Applying insects directly onto cactus infestations.

Evaluation

- **Monitoring:** Regular surveys to assess the reduction in cactus density.

2. *Neochetina eichhorniae* for Water Hyacinth Control

Background: *Neochetina eichhorniae* is used to control water hyacinth (*Eichhornia crassipes*). The weevil larvae tunnel into the petioles, causing significant damage.

Mass Production Techniques

- **Insectaries:** Dedicated facilities for rearing weevils.

Formulation

- **Encapsulation:** Protects weevils during transport and release.

Economics

- **Production Costs:** Initial investment in insectaries offset by long-term control benefits.

Field Release/Application:

- **Inoculative Release:** Establishing self-sustaining populations in infested areas.

Evaluation

- **Effectiveness Metrics:** Measuring reductions in water hyacinth coverage.

Table 1: Mass Production Techniques

S.No	Technique	Description
1	Laboratory Rearing	Controlled environment rearing with optimized diet provision. Maintaining specific temperature, humidity, and light conditions to simulate natural habitats.
2	Field Collection	Identifying and collecting natural populations for augmentation. Using traps, sweep nets, and hand-picking methods. Enhancing field populations by periodically releasing lab-reared individuals.
3	Insectaries	Establishing dedicated facilities for rearing bio-control agents. Equipped with isolation units to prevent contamination and cross-breeding. Maintaining optimal conditions and high-quality diets.

Table 2: Formulation Types

S.No	Formulation	Description
1	Liquid Suspension	Agents suspended in a liquid medium for easy application using standard spraying equipment. Ensures even distribution over large areas. Suitable for foliage and water surface applications.
2	Granular Formulation	Agents mixed with granular carriers (e.g., clay or sand) for soil application. Provides controlled release over time. Ideal for targeting soil-dwelling weed seeds or roots.
3	Encapsulation	Agents enclosed in a protective coating (e.g., alginate or gelatin). Protects from environmental stress and improves shelf-life. Allows controlled release. Beneficial for sensitive agents.
4	Freeze-dried Powders	Dehydrated agents in powder form with a long shelf-life. Easy to store and transport. Can be rehydrated before application, ensuring active and effective agents. Useful for remote areas.

Table 3: Evaluation Metrics

S.No	Metric	Description
1	Weed Suppression	Reduction in weed biomass and coverage. Regular surveys and data collection to measure impact on target weed populations.
2	Agent Performance	Survival, reproduction, and dispersal of bio-control agents. Monitoring agent populations to ensure effective establishment and performance.
3	Environmental Impact	Assessment of non-target effects and changes in biodiversity. Ensuring that bio-control agents do not negatively impact non-target species or ecosystems.

Mass Production of Rice Moth, *Corcyra cephalonica* (Laboratory Host)

Methodology for Rearing of *C. cephalonica*

1. **Sterilization of Rearing Containers**
   - **Wooden Boxes:** Sterilize in a hot air oven at 120°C for 30 minutes.

- **Plastic Trays:** Wash thoroughly with soap and water, then rinse with a mild disinfectant solution to ensure cleanliness and eliminate any contaminants.

2. Preparation of Rearing Medium

- **Ingredients**
  - Heat sterilized broken sorghum grain: 2.5 kg
  - Roasted groundnut powder: 100 grams
  - Yeast: 5 grams
  - Wettable sulfur: 5 grams
  - Streptomycin sulfate: 0.05 grams
- **Procedure**
  - Mix all the ingredients thoroughly in each sterilized box or tray to create a uniform rearing medium.
  - Ensure that the mixture is free from lumps to facilitate easy feeding for the larvae.

3. Inoculation of *Corcyra* Eggs

- **Egg Sprinkling:** Sprinkle *Corcyra* eggs at a rate of 1 cc per box or tray. Ensure even distribution of eggs across the surface of the rearing medium.
- **Covering:** Cover the boxes or trays with lids to prevent contamination and escape of larvae.
- **Labeling:** Clearly label each box or tray with the date of inoculation to monitor the development stages accurately.

4. Developmental Stages

- **Larval Stage**
  - **Feeding:** The hatching larvae feed on the grain by webbing it together. Ensure the medium is not too dry or too moist to promote healthy larval growth.
  - **Duration:** The larval period lasts for 30-35 days.
- **Pupal Stage**
  - **Pupation Site:** Pupation occurs inside the web created by the larvae.
  - **Duration:** The pupal period lasts for 5-7 days.
- **Adult Emergence**
  - **Timeline:** Adult moths emerge after 30-45 days from the date of egg inoculation.

- **Collection:** Collect emerging adults every morning and transfer them to a specially designed mating drum made of G.I. with a wire mesh bottom.

5. Mating and Egg Collection

- **Mating Drum:** Place adults in the mating drum and provide cotton soaked with 20% honey + vitamin E solution as food.
- **Egg Collection**
  - Place a blotting paper in a tray at the bottom of the mating drum to collect the eggs.
  - Pour the eggs into a container, tilting slightly downward to separate eggs from dust particles.
  - Clean the eggs further by passing them through different sized sieves (10, 15, and 40 meshes) to remove impurities.

6. Utilization and Storage of Eggs

- **For Production:** Utilize the cleaned *Corcyra* eggs for *Trichogramma* production or as a host culture for other biological control agents.
- **Storage:** If immediate use is not required, store the eggs in a refrigerator at 10°C for up to 7 days.

7. Optimal Rearing Conditions

- **Temperature:** Maintain a temperature of 28±2°C.
- **Relative Humidity:** Maintain relative humidity at 75±5%.

Flowchart for Mass Multiplication of *C. cephalonica*

Below is a flowchart summarizing the process for mass multiplication of *C. cephalonica*:

1. **Sterilization of Containers**
   - Wooden Boxes: Hot air oven at 120°C for 30 minutes
   - Plastic Trays: Wash with soap and water, disinfect
2. **Preparation of Rearing Medium**
   - 2.5 kg broken sorghum grain (heat sterilized)
   - 100 grams roasted groundnut powder
   - 5 grams yeast
   - 5 grams wettable sulfur
   - 0.05 grams streptomycin sulfate

3. **Inoculation of Eggs**
   - Sprinkle 1 cc *Corcyra* eggs per box or tray
   - Cover with lid
   - Label with inoculation date
4. **Developmental Stages**
   - Larval Stage: 30-35 days, feed on grain by webbing
   - Pupal Stage: 5-7 days, pupate inside the web
   - Adult Emergence: 30-45 days post-inoculation
5. **Collection of Adults**
   - Collect adults every morning
   - Transfer to mating drum with wire mesh bottom
6. **Mating and Egg Collection**
   - Provide 20% honey + vitamin E solution
   - Collect eggs on blotting paper
   - Clean eggs with sieves (10, 15, 40 meshes)
7. **Utilization and Storage**
   - Use for *Trichogramma* production or host culture
   - Store eggs at 10°C for up to 7 days
8. **Optimal Rearing Conditions**
   - Temperature: 28±2°C
   - Relative Humidity: 75±5%

Table 1: Ingredients for Rearing Medium

S.No	Ingredient	Quantity (per box/tray)
1	Broken Sorghum Grain	2.5 kg (heat sterilized)
2	Roasted Groundnut Powder	100 grams
3	Yeast	5 grams
4	Wettable Sulfur	5 grams
5	Streptomycin Sulfate	0.05 grams

Table 2: Developmental Stages of *C. cephalonica*

S.No	Stage	Duration	Description
1	Egg	4-7 days	Hatching into larvae
2	Larval Stage	30-35 days	Feeding on grain, webbing
3	Pupal Stage	5-7 days	Pupation inside web
4	Adult Emergence	30-45 days from egg	Emerging from pupae, ready for collection

Table 3: Optimal Rearing Conditions

S.No	Condition	Optimal Range
1	Temperature	28±2°C
2	Relative Humidity	75±5%

Activity: Preparing a Flowchart for Mass Multiplication of *C. cephalonica* and Recording Life Stages

Flowchart

1. Sterilization of Containers
   - Hot air oven (wooden boxes)
   - Washing and disinfecting (plastic trays)
2. Preparation of Rearing Medium
   - Mixing broken sorghum grain, groundnut powder, yeast, wettable sulfur, streptomycin sulfate
3. Inoculation of Eggs
   - Sprinkling eggs
   - Covering and labeling
4. Developmental Stages
   - Feeding and webbing (larval stage)
   - Pupation inside web (pupal stage)
   - Adult emergence
5. Collection of Adults
   - Daily morning collection
   - Transfer to mating drum
6. Mating and Egg Collection
   - Providing food (honey + vitamin E)
   - Collecting and cleaning eggs

7. Utilization and Storage of Eggs
    - Using for production or host culture
    - Storing at 10°C
8. Maintaining Optimal Conditions
    - Temperature and humidity control

Recording Life Stages

- **Egg Stage:** Record hatching time and egg viability.
- **Larval Stage:** Monitor feeding patterns, webbing extent, and larval health.
- **Pupal Stage:** Track pupation sites, duration, and pupal health.
- **Adult Emergence:** Record emergence rates, adult health, and mating behavior.

Mass Production of *Chelonus blackburni* (Egg-Larval Parasitoid)
Materials Required

- Polythene bags
- Rubber bands
- Scissors
- Gum
- Brush
- Tea strainer
- 50% honey solution
- Card (5 x 5 cm)
- Refrigerator
- UV lamp
- Plastic containers (1.5 liters) with plastic mesh windows for aeration
- Cotton swabs
- Sterilized cumbu medium (500 g)

Methodology for Mass Production of *Chelonus blackburni*

1. **Preparation of Egg Cards**
    - Paste a set of 100, 0-24 hr old *Corcyra* eggs (not exposed to UV) onto a 5 x 5 cm card. Use gum to ensure the eggs stick firmly to the card.
    - Ensure the eggs are spread evenly across the card to facilitate parasitization.

2. **Exposure to *Chelonus blackburni***
    - Place the egg card into a 1.5-liter plastic container containing 30°C. *blackburni* adults.
    - The plastic container should have windows with plastic mesh for proper aeration.
    - Place two cotton swabs inside the container:
    - One soaked in 10% honey solution to provide nutrition for the adult parasitoids.
    - The other soaked in drinking water to maintain humidity.
    - Close the container with a cloth-covered cotton plug to prevent the escape of the parasitoids while allowing air exchange.
3. **Post-Exposure Handling:**
    - After 24 hours of exposure, remove the egg card from the container.
    - Place the exposed egg card onto 500 g of sterilized cumbu medium. The medium should be spread evenly in a rearing box or tray to provide a suitable environment for the developing larvae.
4. **Development and Emergence**
    - Over the next 30 days, *Chelonus blackburni* larvae develop within the *Corcyra* eggs and subsequently in the cumbu medium.
    - The parasitoids form cocoons in the medium, from which adults will emerge.
    - Monitor the development stages, noting the changes in the parasitized eggs and larvae.
5. **Collection of Adults**
    - After 30 days, adult parasitoids start emerging from the cocoons.
    - Collect the emerging adults daily. Adults live for approximately 25 days and can produce about 400 eggs each during their lifespan.

Field Release

- The collected *Chelonus blackburni* adults are taken to the fields for release.
- Recommended release rate:
    - 1 parasitoid per plant
    - Or 8000 parasitoids per acre

Activities: Flowchart and Observation Record

Flowchart for Mass Production of *Chelonus blackburni*

1. **Preparation of Egg Cards**
   - Paste 100 *Corcyra* eggs on 5 x 5 cm card
   - Use gum for secure attachment
2. **Exposure to Parasitoids**
   - Place egg card in 1.5-liter plastic container with 30 *C. blackburni* adults
   - Ensure aeration with plastic mesh windows
   - Insert cotton swabs soaked in 10% honey solution and drinking water
   - Close with cloth-covered cotton plug
3. **Post-Exposure Handling:**
   - Remove egg card after 24 hours
   - Place egg card on 500 g sterilized cumbu medium
4. **Development and Emergence**
   - Monitor development for 30 days
   - Collect emerging adults daily
5. **Field Release**
   - Release 1 parasitoid per plant or 8000 per acre

Table: Developmental Stages and Observations

S.No	Stage	Duration	Observations
1	Egg Parasitization	24 hours	Egg card exposed to *C. blackburni*
2	Larval Development	30 days	Larvae develop in *Corcyra* eggs
3	Pupal Stage	30 days	Cocoons form in cumbu medium
4	Adult Emergence	30 days	Adults emerge from cocoons
5	Adult Collection	Daily	Collect adults, lifespan ~25 days
6	Field Release	N/A	Release 1 parasitoid/plant or 8000/acre

Mass Production of *Trichogramma* Species (Egg Parasitoid)

Methodology

Preparation of *Corcyra* Eggs

1. **Cleaning**
   - Clean fresh *Corcyra* eggs by passing them through 15, 30, and 45 mesh sieves to remove debris and ensure uniform size.

2. **Preparation of "Trichocard"**
   - Cut cardboard sheets into 10 x 10 cm cards, which can accommodate 1 cc of eggs.
   - Apply a thin layer of gum uniformly on the card.
   - Sprinkle the cleaned *Corcyra* eggs uniformly over the gum-coated card.
   - Remove excess eggs from the card using a brush.
   - Allow the card to shade dry for 30 minutes.
3. **UV Treatment**
   - Treat the egg cards under a UV lamp for 30 minutes to sterilize the eggs and prevent premature hatching.

Parasitization Process

4. **Insertion into Polythene Bags**
   - Take a polythene bag and insert the UV-treated "Trichocard" along with a nucleus card in the ratio of 6:1 (6 *Corcyra* egg cards: 1 *Trichogramma* nucleus card).
   - Provide a cotton swab soaked in 50% honey and vitamin E solution inside the bag to nourish the adult *Trichogramma*.
5. **Parasitization**
   - Remove the *Trichocard* after 2 days.
   - Observe the *Corcyra* eggs; they turn black on the 3rd day, indicating successful parasitization by *Trichogramma*.

Field Release and Storage

6. **Field Release**
   - The parasitized egg cards should be released immediately in the fields.
   - If not released immediately, store them in a refrigerator at 10°C for up to 21 days.
   - For field release, place, tie, or staple the parasitized cards on the leaf sheath of the plant.
   - *Trichogramma* parasitoids emerge 7 days after parasitization at room temperature.
   - If cold-stored, take out the cards and keep them at room temperature for a day before releasing them into the field.

- Cut the egg cards into smaller sections along the lines and staple them on the plant.

Production of Trichocard Under Laboratory Conditions

Materials Required

- Cardboard sheets (10 x 10 cm)
- Gum
- Brush
- UV lamp
- Polythene bags
- Cotton swabs
- 50% honey solution with vitamin E

Activities: Flowchart and Observation Record

Flowchart for Mass Production of *Trichogramma chilonis*

1. **Cleaning and Preparation of Eggs**
   - Clean *Corcyra* eggs using mesh sieves (15, 30, 45 meshes).
   - Prepare "Trichocard" by cutting cardboard sheets (10 x 10 cm).
   - Apply gum and sprinkle cleaned eggs.
   - Shade dry for 30 minutes.
2. **UV Treatment**
   - Treat egg cards under UV lamp for 30 minutes.
3. **Parasitization Process:**
   - Insert UV-treated "Trichocard" and nucleus card into polythene bag (6:1 ratio).
   - Provide cotton swab soaked in 50% honey + vitamin E solution.
   - Remove *Trichocard* after 2 days.
4. **Observation**
   - Eggs turn black on the 3rd day, indicating parasitization.
5. **Field Release and Storage**
   - Release parasitized egg cards immediately or store at 10°C up to 21 days.
   - Place, tie, or staple cards on leaf sheath of plants.
   - Parasitoids emerge 7 days after parasitization at room temperature.
   - If cold-stored, keep at room temperature for a day before field release.

Table: Developmental Stages and Observations

S.No	Stage	Duration	Observations
1	Egg Cleaning	Immediate	Cleaning with mesh sieves
2	Trichocard Preparation	30 minutes	Apply gum, sprinkle eggs, shade dry
3	UV Treatment	30 minutes	Sterilization under UV lamp
4	Parasitization	2 days	Insert cards into polythene bags, provide food
5	Observation	3rd day	Eggs turn black, indicating parasitization
6	Field Release/Storage	Up to 21 days	Immediate release or storage at 10°C
7	Parasitoid Emergence	7 days	Emergence at room temperature

Observation Record

- **Egg Cleaning:** Record the quantity of eggs cleaned and any debris removed.
- **Trichocard Preparation:** Note the number of cards prepared and the uniformity of egg distribution.
- **UV Treatment:** Document the duration of exposure and the number of treated cards.
- **Parasitization Process:** Monitor the parasitization success by checking the color change of eggs.
- **Field Release and Storage:** Record the number of parasitized cards released or stored and the emergence rate of *Trichogramma*.

Mass Production of *Bracon brevicornis* (Larval Parasitoid)

Methodology for Mass Production of *Bracon brevicornis*

1. Host Rearing

Bracon brevicornis can be reared in the laboratory using the rice moth, *Corcyra cephalonica*, as an alternate host. The "Sandwich" technique is often used for small-scale culture.

2. Setting up the Culture

1. **Preparation of the Chimney**
    - Use a glass chimney and cover both ends with muslin cloth, securing them with rubber bands.
    - Confine about 20 mated *B. brevicornis* females inside the chimney.
    - Stick a cotton swab soaked in a 50% honey water solution to the side of the chimney to serve as food.

2. Nutrition

- Adult nutrition is crucial as it influences the sex ratio, with a high protein diet potentially increasing the number of female progeny.
- 'Proteinex' or similar high-protein supplements can be used to enhance results.
- Replacing honey with laevulose or fructose may also be beneficial.
- Exposure to sunlight can stimulate mating, oogenesis, and egg fertilization.

3. Parasitization Process

1. **Preparing Host Larvae**
    - Place 10 full-grown *Corcyra* larvae between two sheets of facial tissue paper.
    - Place the tissue over the muslin sheet covering the wider mouth of the chimney.
    - Cover the tissue with another muslin sheet and secure it with rubber bands.
2. **Parasitization**
    - Position the chimney with the host larvae facing a window or light source to attract the female *B. brevicornis*.
    - The females will probe through the muslin and paralyze the larvae, laying about 25 eggs per day on each larva.
    - After 24 hours, remove the tissue sheets with parasitized larvae.
3. **Development and Emergence**
- Hold the tissue sheets in flat plastic containers until the parasitoid grubs hatch, develop, and spin cocoons.
- The lifecycle stages are as follows:
    - Egg stage: 28-36 hours
    - Larval stage: 4-7 days
    - Pupal stage: 3-6 days
    - Adult stage: 15-40 days
4. **Collecting Adults**
    - Female parasitoids can lay 150-200 eggs in their lifetime.
    - Emerging adults are collected daily for mating and further egg laying.
    - Adults survive for up to 15-40 days, with egg laying tapering off after the first ten days.
    - Two-day-old adults can be stored for 30 days at 5°C and 50-60% RH.

Field Release

- *Bracon hebetor* is released at a rate of 8000 adults per acre for controlling cotton bollworm.
- *Bracon brevicornis* is released at a rate of 10 adults per tree for controlling coconut black-headed caterpillar.

Flowchart for Mass Production of *Bracon brevicornis*

1. **Preparation of Chimney**
   - Use a glass chimney, cover both ends with muslin cloth, and secure with rubber bands.
   - Confine 20 mated *B. brevicornis* females inside the chimney.
   - Provide a cotton swab soaked in 50% honey water solution.
2. **Host Larvae Preparation**
   - Place 10 full-grown *Corcyra* larvae between two sheets of facial tissue paper.
   - Place the tissue over the muslin sheet on the chimney's wider mouth.
   - Cover with another muslin sheet and secure with rubber bands.
3. **Parasitization**
   - Position the chimney near a window or light source to attract females.
   - Females probe through the muslin, paralyze larvae, and lay eggs.
4. **Post-Parasitization:**
   - After 24 hours, remove tissue sheets with parasitized larvae.
   - Hold in flat plastic containers for development and cocoon formation.
5. **Development Stages**
   - Egg stage: 28-36 hours
   - Larval stage: 4-7 days
   - Pupal stage: 3-6 days
   - Adult stage: 15-40 days
6. **Collecting and Storing Adults**
   - Collect emerging adults daily.
   - Store two-day-old adults at 5°C and 50-60% RH for up to 30 days.

Table: Developmental Stages and Observations

S.No	Stage	Duration	Observations
1	Egg Stage	28-36 hours	Eggs laid on paralyzed *Corcyra* larvae
2	Larval Stage	4-7 days	Parasitoid larvae develop inside host
3	Pupal Stage	3-6 days	Parasitoid pupae form cocoons
4	Adult Emergence	15-40 days	Adults emerge form cocoons
5	Adult Collection	Daily	Collect adults for mating and egg laying
6	Storage	Up to 30 days	Store adults at 5°C and 50-60% RH

Activities: Observation Record

- **Host Larvae Preparation:** Record the number of *Corcyra* larvae used and the condition of the tissue sheets.
- **Parasitization:** Monitor the number of eggs laid per larva and the behavior of *B. brevicornis* females.
- **Development Stages:** Document the duration of each developmental stage and any abnormalities observed.
- **Adult Collection:** Note the number of adults collected daily and their survival rate.
- **Field Release:** Record the number of adults released per acre/tree and the observed impact on target pests.

Mass Production of *Brachymeria* Species (Pupal Parasitoid)

Production Procedure

1. Setting Up the Culture

1. **Initial Release**
    - Release 50 adults of *Brachymeria nosatoi* (both sexes) in a clean, dry cylindrical jar measuring 17.5 cm in height and 6.75 cm in diameter.
    - Insert a 12 cm long and 6.25 cm wide piece of cardboard into the jar to facilitate the parasitoids' movement and provide a resting surface.
2. **Securing the Jar**
    - Secure the mouth of the jar with a piece of muslin cloth, tightened with rubber bands.
    - Place the jar horizontally to allow better access and movement for the parasitoids.
3. **Maintenance and Transfer**
    - Transfer the parasitoids to a fresh, clean jar every 4 to 5 days to maintain hygiene and reduce stress on the insects.

- Provide undiluted honey daily in minute droplets on wax-coated paper for adult nutrition.

2. Stimulating Mating

4. Sunlight Exposure

- Place the jar containing the parasitoids in diffused sunlight for 10-15 minutes daily for about 3-4 days.
- Exposure to sunlight stimulates mating, enhancing reproductive success.

3. Preparing Host Pupae

5. Handling Pupae

- Carefully remove the pupae of *Opisina arenosella* with their cocoons and silken galleries intact. Alternatively, use leaf bits containing pupae within cocoons and silken galleries.
- Place the pupae on a 12 x 6 cm piece of cardboard, ensuring they are accessible to the parasitoids from all three sides.

6. Parasitization Setup

- Insert the cardboard piece containing the pupae into the horizontally placed jar with the mated parasitoids for parasitization.
- Ensure the pupae are arranged so that the parasitoids can easily access them.

4. Parasitization Process

7. Oviposition

- The parasitoids disorganize the pupal tissues by standing on the galleries and repeatedly thrusting their ovipositors into the pupae to oviposit.
- Place the pupae without cocoons and silken galleries on the cardboard and cover them with the silken galleries to facilitate parasitization. The parasitoids will not parasitize naked pupae.

8. Exposure Duration

- Depending on the activity level of the female parasitoids, expose the host pupae for a period of 4-6 hours for effective parasitization.

Field Release

- Release *Brachymeria nosatoi* at a rate of 20 adults per tree to control the coconut black-headed caterpillar.

Activities: Observation and Recording

Observations

1. **Initial Setup**
    - Record the number of adults released and the initial setup conditions.
2. **Mating and Nutrition**
    - Monitor the behavior and mating success of the parasitoids during the sunlight exposure period.
    - Record the frequency and quantity of honey provided.
3. **Parasitization Process**
    - Observe the interaction between the parasitoids and the host pupae.
    - Record the number of pupae parasitized and the duration of exposure.
4. **Development and Emergence**
    - Monitor the development stages of the parasitized pupae until the emergence of adult parasitoids.
    - Note any changes in the pupae, including discoloration or structural changes.

Flowchart for Mass Production of *Brachymeria nosatoi*

1. **Setting Up the Culture**
    - Release 50 adults (both sexes) in a cylindrical jar (17.5 x 6.75 cm).
    - Insert a 12 x 6.25 cm cardboard piece for movement and resting.
    - Secure with muslin cloth and rubber bands, place jar horizontally.
2. **Maintenance and Transfer**
    - Transfer parasitoids to a fresh jar every 4-5 days.
    - Provide undiluted honey daily on wax-coated paper.
3. **Stimulating Mating**
    - Expose jar to diffused sunlight for 10-15 minutes daily for 3-4 days.
4. **Preparing Host Pupae**
    - Remove pupae of *O. arenosella* with cocoons and silken galleries intact.
    - Place on a 12 x 6 cm cardboard piece.
5. **Parasitization Setup**
    - Insert cardboard with pupae into the jar with parasitoids.
    - Ensure pupae are accessible from all sides.

6. **Parasitization Process**
   - Parasitize pupae for 4-6 hours.
   - Record changes in pupae and parasitoid activity.

Table: Developmental Stages and Observations

S.No	Stage	Duration	Observations
1	Initial Setup	Immediate	Number of adults released, initial setup
2	Mating Stimulus	3-4 days	Sunlight exposure, mating behavior
3	Parasitization	4-6 hours	Interaction with host pupae, oviposition
4	Development	Continuous	Monitoring development stages
5	Emergence	15-40 days	Emergence of adult parasitoids
6	Field Release	Immediate	Release rate: 20 adults/tree

Mass Production of *Bracon brevicornis* (Larval Parasitoid)

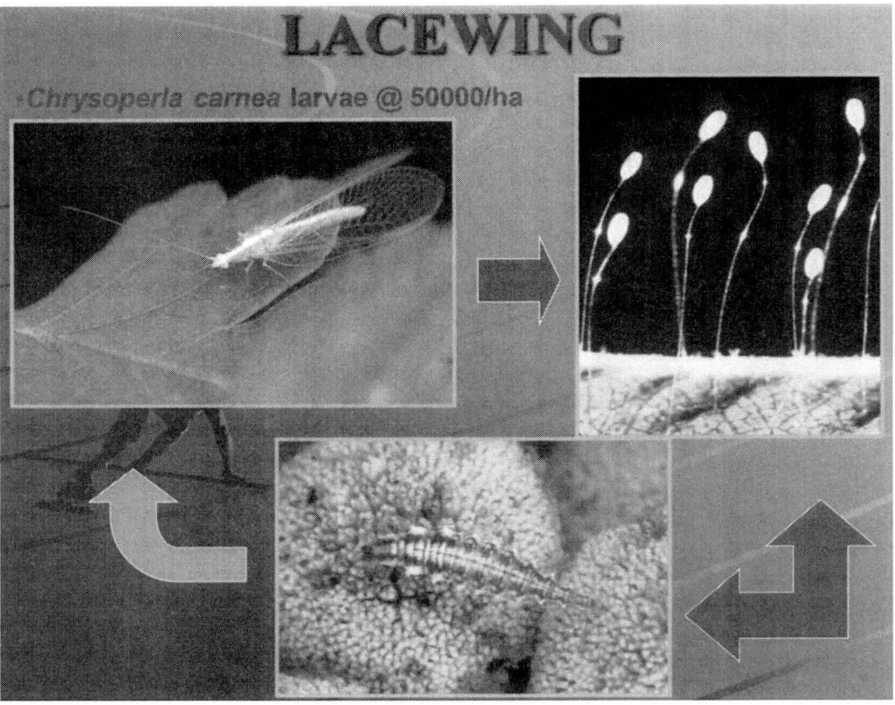

Production Procedure

Bracon brevicornis is amenable for mass rearing in the laboratory using the alternate host, *Corcyra cephalonica*. For small-scale culture, the "Sandwich" technique is adequate.

1. Initial Setup

1. Confine Mated Females

- Confine about 20 mated females of *B. brevicornis* in a glass chimney.
- Cover both sides of the chimney with muslin cloth secured with rubber bands.

2. Nutrition

- Stick a cotton swab soaked in 50% honey water solution to the side of the chimney to serve as food.
- Proper adult nutrition is crucial as it influences the sex ratio. High protein diets can improve the sex ratio, producing more female progeny. 'Proteinex' can be used to achieve these results.
- Replacing honey with laevulose or fructose is beneficial in some cases.
- Exposure to sunlight frequently stimulates mating, oogenesis, and egg fertilization.

2. Parasitization Process

3. Preparation of Host Larvae

- Place about 10 full-grown larvae of *Corcyra cephalonica* between two sheets of facial tissue paper.
- Place the tissue over the muslin sheet covering the wider mouth of the chimney.
- Cover the tissue again with a sheet of muslin and fasten it with a pair of rubber bands.

4. Exposure to Light

- Place the chimney with the host larvae facing a window or light source. This setup attracts the female *B. brevicornis* to the host larvae.
- Females probe through the muslin, paralyze the larvae, and lay about 25 eggs per day on each larva.

5. Post-Parasitization Handling:

- After 24 hours, remove the tissue sheets bearing parasitized larvae.
- Place the parasitized larvae in flat plastic containers until the parasitoid grubs hatch, complete development, and spin cocoons.

3. Development and Collection

6. Lifecycle Stages

- **Egg Stage:** Completed in 28-36 hours.

- **Larval Stage:** Lasts for 4-7 days.
- **Pupal Stage:** Completed in 3-6 days.
- **Adult Stage:** Adults live for 15-40 days.

7. **Adult Collection**
   - Female *B. brevicornis* can deposit 150-200 eggs in their lifetime.
   - Collect emerging adults for mating and egg laying. Adults usually taper off egg laying after the first ten days.
   - Two-day-old adults can be stored for 30 days at 5°C and 50-60% relative humidity (RH).

Field Release

- *Bracon hebetor* is released at a rate of 8000 adults per acre for cotton bollworm control.
- *Bracon brevicornis* is released at a rate of 10 adults per tree for controlling the coconut black-headed caterpillar.

Activities: Flowchart and Observation Record

Flowchart for Mass Production of *Bracon brevicornis*

1. **Initial Setup**
   - Confine 20 mated females in a glass chimney.
   - Cover with muslin cloth and secure with rubber bands.
   - Provide cotton swab with 50% honey water solution.
2. **Host Larvae Preparation**
   - Place 10 full-grown *Corcyra cephalonica* larvae between facial tissue sheets.
   - Place tissue over muslin sheet on chimney's wider mouth.
   - Cover with another muslin sheet and secure.
3. **Exposure to Light**
   - Position chimney near a window or light source.
   - Females paralyze larvae and lay eggs.
4. **Post-Parasitization**
   - Remove tissue sheets after 24 hours.
   - Place parasitized larvae in flat plastic containers.
5. **Development Stages**
   - Egg stage: 28-36 hours
   - Larval stage: 4-7 days

- Pupal stage: 3-6 days
- Adult stage: 15-40 days

6. **Collection and Storage**
   - Collect emerging adults daily.
   - Store two-day-old adults at 5°C and 50-60% RH for up to 30 days.

Table: Developmental Stages and Observations

S.No	Stage	Duration	Observations
1	Egg Stage	28-36 hours	Eggs laid on paralyzed *Corcyra* larvae
2	Larval Stage	4-7 days	Parasitoid larvae develop inside host
3	Pupal Stage	3-6 days	Parasitoid pupae form cocoons
4	Adult Emergence	15-40 days	Adults emerge from cocoons
5	Adult Collection	Daily	Collect adults for mating and egg laying
6	Storage	Up to 30 days	Store adults at 5°C and 50-60% RH

Observation Record

1. **Initial Setup:** Record the number of adults released and initial setup conditions.
2. **Mating and Nutrition:** Monitor the behavior and mating success of parasitoids during sunlight exposure.
3. **Parasitization Process:** Observe the interaction between parasitoids and host larvae. Record the number of larvae parasitized and duration of exposure.
4. **Development and Emergence:** Monitor the development stages of parasitized larvae until the emergence of adult parasitoids. Note any changes in larvae, including discoloration or structural changes.
5. **Field Release:** Record the number of adults released per acre/tree and observed impact on target pests.

Mass Production of *Chrysoperla carnea* (Predator)

Methodology

Objective: The goal is to establish a sustainable and efficient process for mass rearing *Chrysoperla carnea* to be used as a biological control agent against various pests in agricultural fields.

1. Rearing Environment and Adult Collection

1. **Adult Feeding**

- Adults are provided with a protein-rich diet that consists of yeast, fructose, honey, Proteinex, and water in the ratio of 1:1:1:1. This semisolid paste is placed in three spots on the cloth covering the rearing troughs.

- Place three bits of foam sponge (2 sq.in) soaked in water on the cloth outside the trough to maintain humidity.

2. **Adult Collection and Housing:**
   - Collect adult *Chrysoperla carnea* daily and transfer them to pneumatic glass troughs or G.I. round troughs measuring 30 cm in diameter and 12 cm in height.
   - Line the inside of the troughs with brown paper sheets, which serve as egg-laying substrates.
   - Allow about 250 adults (with a 60% female ratio) into each trough.
   - Cover the troughs with white nylon or georgette cloth secured by rubber bands.

3. **Egg Collection**
   - Adults lay eggs on the brown paper sheets inside the rearing troughs.
   - Collect the adults daily and transfer them to fresh rearing troughs with new food and water supplies.
   - Remove the brown paper sheets containing eggs from the old troughs.

2. Storage and Destalking of Eggs

4. **Storage of Egg Sheets**
   - Store the brown paper sheets with *Chrysoperla* eggs at 10°C in a B.O.D. incubator or refrigerator for up to 21 days.
   - When needed for culturing or field release, keep the egg sheets at room temperature for a day. During this period, the eggs will turn brown and hatch approximately three days later.

5. **Destalking and Hatching**
   - Monitor the hatching process and ensure that the first instar larvae are either utilized for further culturing or prepared for field release.

3. Field Release

6. **Preparation for Release**
   - Prepare the first instar larvae for field release by placing them in plastic containers with 1-2 cc of *Corcyra* eggs and loose paper strips.

7. **Release in Fields**
   - Release *Chrysoperla carnea* larvae in cotton fields at a rate of 20,000 to 40,000 larvae per acre.

- Conduct releases 3-5 times at 10-day intervals to control pests such as aphids, whitefly, *Spodoptera*, *Heliothis*, pink bollworm, thrips, and mites.
- While walking across the fields, drop the paper strips with larvae randomly to ensure even distribution.

Flowchart for Mass Production of *Chrysoperla carnea*

1. **Adult Rearing**
   - Provide protein-rich diet (yeast, fructose, honey, Proteinex, water in 1:1:1:1 ratio).
   - Maintain humidity with water-soaked foam sponges.
   - House 250 adults (60% females) in rearing troughs lined with brown paper.
2. **Egg Collection**
   - Collect adults daily, transfer to fresh troughs.
   - Remove brown paper sheets with eggs from old troughs.
3. **Storage and Hatching**
   - Store egg sheets at 10°C for up to 21 days.
   - Keep at room temperature for a day before use.
   - Hatch larvae for further culturing or field release.
4. **Field Release**
   - Place larvae in plastic containers with *Corcyra* eggs and paper strips.
   - Release 20,000 to 40,000 larvae/acre, 3-5 times at 10-day intervals.
   - Randomly distribute paper strips with larvae in fields.

Table: Developmental Stages and Observations

S.No	Stage	Duration	Observations
1	Adult Rearing	Daily	Collect adults, provide diet and water
2	Egg Laying	Continuous	Eggs laid on brown paper sheets
3	Egg Storage	Up to 21 days	Store at 10°C, monitor for hatching
4	Hatching	1-3 days	Eggs turn brown, hatch into larvae
5	Field Release	Every 10 days	Release 20,000 to 40,000 larvae/acre

Observation Record

1. **Adult Rearing**
   - Record the number of adults introduced into rearing troughs.
   - Monitor feeding and water supply.

2. **Egg Collection and Storage**
    - Track the number of eggs collected on brown paper sheets.
    - Document storage conditions and duration.
3. **Hatching and Larvae Collection**
    - Observe egg color changes and hatching rates.
    - Record the number of first instar larvae hatched.
4. **Field Release**
    - Note the number of larvae released and the frequency of releases.
    - Monitor pest control effectiveness in the field.

Mass Production of *Cryptolaemus montrouzieri* (Predator)

Methodology for Mass Production of *Cryptolaemus montrouzieri*

1. Colony Establishment

- **Collection**
    - Collect colonies of mealy bugs from fields, focusing on guava plantations, vineyards, papaya, citrus, and pomegranate gardens as these are good reservoirs.
    - Purify the mealy bug colonies in the laboratory to obtain populations free from parasitoids and scavenging ants.

2. Culture Maintenance

- **Host Preparation**
    - Use pumpkins (red) as the culture medium for mealy bugs as maintaining colonies on natural host plants is challenging.
    - Select fleshy pumpkins with intact peduncle, deep ridges, and furrows, weighing approximately 2.5 kg. Ensure they are free from wounds and mouldy patches.
    - Soak the pumpkins in a 0.5% carbendazim solution for 1 minute and then shade dry them. Plug any cut ends and wounds with molten wax to prevent infections and dehydration.
    - Provide pieces of paper around the pumpkins to facilitate movement and breeding of mealy bugs.
- **Setup**
    - Place the prepared pumpkins in large-sized cages positioned over stainless steel stands.
    - Ensure the cages are ant-proof as mealy bugs secrete honeydew, which attracts ants. Use barriers or ant traps as necessary.

- Collect the ovisacs of healthy adult mealy bugs and place them individually on fresh pumpkins in the laboratory. Allow the eggs to hatch and multiply.

- **Maintenance**
  - In about a month, the mealy bugs will cover the entire surface of the pumpkins. Use this stock to establish subsequent colonies.
  - During active growth periods, collect ovisacs with the help of a camel hairbrush and transfer them to fresh pumpkins prepared in the same manner.
  - Sterilize cages and stainless steel ware using common bleach to prevent fungal invasion. Dispose of pumpkins showing mould symptoms immediately.

3. Mass Production

- **Rearing *Cryptolaemus montrouzieri***
  - After 25 days of releasing the mealy bugs onto the pumpkins, introduce 10 mated adult females of *Cryptolaemus montrouzieri* into the cage.
  - To facilitate the pupation of grubs, place pieces of paper on the bottom of the cage.
  - After 1.5 to 2 months, collect the emerging beetles daily in glass vials for up to 5-10 days.

Flowchart for Mass Production of *Cryptolaemus montrouzieri*

1. **Colony Establishment**
   - Collect mealy bug colonies from fields.
   - Purify in the lab to remove parasitoids and ants.
2. **Culture Maintenance**
   - Prepare pumpkins (red) for mealy bug culture.
   - Soak in 0.5% carbendazim solution for 1 minute and shade dry.
   - Place pumpkins in ant-proof cages over stainless steel stands.
   - Collect and transfer ovisacs to fresh pumpkins.
3. **Mass Production**
   - After 25 days of mealy bug release, introduce 10 mated adult *Cryptolaemus montrouzieri* females.
   - Place paper pieces for grub pupation.

- Collect emerging beetles in glass vials daily for 5-10 days after 1.5 to 2 months.

Table: Developmental Stages and Observations

S.No	Stage	Duration	Observations
1	Colony Establishment	Initial setup	Collection and purification of mealy bugs
2	Culture Maintenance	Continuous	Pumpkin preparation, mealy bug rearing
3	Mealy Bug Development	1 month	Mealy bugs cover pumpkin surface
4	Parasitization Introduction	25 days	Introduction of *Cryptolaemus* females
5	Pupation and Beetle Emergence	1.5-2 months	Collection of emerging beetles daily

Observation Record

1. **Colony Establishment**
    - Record the source and quantity of mealy bug colonies collected.
    - Monitor and document the purification process.
2. **Culture Maintenance**
    - Note the number of pumpkins prepared and their condition.
    - Record the ovisacs collected and their transfer to fresh pumpkins.
    - Monitor for fungal invasions and document sterilization procedures.
3. **Mass Production**
    - Record the date of mealy bug release and introduction of *Cryptolaemus montrouzieri* females.
    - Monitor and document the pupation process and the emergence of beetles.
    - Collect data on the number of beetles emerged and their collection dates.

Mass Production of *Zygogramma bicolorata* (Phytophagous Natural Enemy of Weed)

Methodology

1. Adult Setup and Egg Collection

1. **Initial Setup**
    - Place 10 pairs of adult *Zygogramma bicolorata* (male and female) on bouquets of *Parthenium* leaves in a 14 x 12 cm transparent plastic container.

- Monitor for egg laying and replace the leaves with fresh bouquets once eggs are observed.

2. **Continuous Replacement**
   - Repeat the process of replacing bouquets in the egg-laying jars for one month to ensure continuous egg collection.

2. Plant Preparation

3. **Transplanting *Parthenium* Plants**
   - Remove small *Parthenium* plants from soil and transplant them into 45 x 60 x 90 cm cages with zinc sheet trays at the bottom.
   - Fill these trays with soil, transplant the *Parthenium* plants, and water them daily to ensure healthy growth.

3. Egg Transfer and Larval Development

4. **Egg Transfer**
   - Once the *Parthenium* plants start growing, place leaves with *Zygogramma bicolorata* eggs over them. Transfer around 100-150 eggs into each cage.
5. **Larval Feeding and Pupation**
   - Upon hatching, the larvae will feed on the leaves and pupate in the soil.
   - Ensure the larvae consume all the plants inside the cages. Generally, the ratio is 10-15 grubs per plant.
   - Provide around 15-20 small *Parthenium* plants in each cage to sustain the larvae until pupation.
6. **Adult Emergence**
   - Each cage can yield around 100-125 adults.
   - The process can also be adapted for open field breeding by covering *Parthenium* plants with walk-in field cages and releasing 1 pair of adults per 2 plants.

Activities

1. **Collecting and Rearing *Zygogramma bicolorata*:**
   - Collect *Zygogramma bicolorata* adults and rear them in the laboratory using the described methodology.
2. **Illustrating the Life Cycle**
   - Document the life cycle stages of *Zygogramma bicolorata* from egg, grub, and adult stages.

- Note and describe their feeding behavior on *Parthenium* leaves.

Flowchart for Mass Production of *Zygogramma bicolorata*

1. **Adult Setup**
   - Place 10 pairs of adults on *Parthenium* leaf bouquets in plastic containers.
   - Replace leaves with fresh bouquets once eggs are observed.
   - Repeat for one month.
2. **Plant Preparation**
   - Transplant small *Parthenium* plants into 45 x 60 x 90 cm cages.
   - Water daily and ensure healthy growth.
3. **Egg Transfer**
   - Place leaves with *Zygogramma* eggs on growing *Parthenium* plants.
   - Transfer 100-150 eggs per cage.
4. **Larval Development**
   - Larvae feed on leaves and pupate in the soil.
   - Provide 15-20 *Parthenium* plants per cage.
   - Cages yield around 100-125 adults.
5. **Field Breeding (Optional)**
   - Cover *Parthenium* plants with walk-in field cages.
   - Release 1 pair of adults per 2 plants.

Table: Developmental Stages and Observations

S.No	Stage	Duration	Observations
1	Egg Laying	Continuous	Eggs laid on *Parthenium* leaves
2	Egg Incubation	5-10 days	Eggs hatch into larvae
3	Larval Feeding	2-3 weeks	Larvae feed on leaves, consume plants
4	Pupation	7-10 days	Larvae pupate in soil
5	Adult Emergence	1-2 weeks	Adults emerge from soil
6	Adult Collection	Daily	Collect adults for further rearing
7	Field Release (Optional)	N/A	Release adults in field cages

Observation Record

1. **Egg Collection**
   - Monitor and document the number of eggs laid daily.
   - Record the frequency of leaf replacement.
2. **Plant Preparation**

- Document the preparation process and health status of transplanted *Parthenium* plants.

3. **Larval Development**
   - Observe and record the feeding behavior and growth rate of larvae.
   - Note the number of larvae per plant and their consumption rate.
4. **Adult Emergence**
   - Track the emergence rate and number of adults collected daily.
   - Record any anomalies or issues in the development process.

Mass Production of *Neochetina* Species (Natural Enemies of Weeds)

Methodology

Species

- *Neochetina eichhorniae* and *Neochetina bruchi*

Materials Required

- Water hyacinth plants
- Outdoor tanks ($4m^2$ x 1 m depth) cement cisterns or plastic fishing pools (120 cm diameter x 60 cm depth)
- Cow dung (200 g/m^3)
- Superphosphate (40 g/m^3)
- Urea (10 g/m^3)
- Plastic troughs (61 x 40 x 30 cm)

Steps

1. Preparation of Growth Medium

Outdoor Tanks or Pools

- The setup involves outdoor tanks or plastic fishing pools that provide an ideal environment for both the water hyacinth and the *Neochetina* weevils.
- These tanks or pools should be filled with water to a depth of 1 meter.
- Enhance the growth medium by adding cow dung (200 g/m^3), superphosphate (40 g/m^3), and urea (10 g/m^3) to the water. This nutrient-rich environment promotes faster and more prolific growth of water hyacinth plants, which in turn supports the multiplication of *Neochetina* weevils.

2. Initial Exposure in the Laboratory

Laboratory Exposure:

- Start the process indoors by placing about 25 water hyacinth plants in plastic troughs measuring 61 x 40 x 30 cm.
- Release 50 adult *Neochetina* weevils (both *N. eichhorniae* and *N. bruchi*) onto these plants.
- After a week, collect the adults and transfer them to fresh plants to ensure continuous exposure and to build up the population of *Neochetina* weevils.

3. Multiplication Process

Outdoor Setup

- Transfer the exposed plants from the laboratory to the prepared outdoor tanks or plastic fishing pools.
- This step should be done carefully to ensure that the eggs and larvae on the plants are not disturbed.
- Continue the process of exposing fresh plants to *Neochetina* adults until the tanks or pools are filled with infested plants. Each set of adults can be reused for at least 15 exposures before they need to be replaced.

Continuous Multiplication

- After the initial batch of fresh adults starts emerging, typically after 3-4 months, collect them at fortnightly intervals for field release.
- Eggs laid by the adults before collection will produce subsequent generations, ensuring continuous multiplication.
- Regularly add fresh plants to the tanks or pools as older plants start drying out to maintain a steady supply of water hyacinth for the weevils.

4. Collection and Field Release

Adult Collection

- From a 120 cm diameter and 60 cm deep plastic fishing pool, which can accommodate up to 100 plants, between 100-250 adults can be collected per month.
- These adults are then prepared for field release to control water hyacinth infestations.

Field Release

- The collected *Neochetina* adults are released into infested water bodies to manage water hyacinth. The release rate and timing should be planned to maximize the impact on the water hyacinth population, ensuring that the weevils can effectively reduce the infestation.

Life Cycle and Illustration of *Neochetina bruchi*

1. Egg Stage

- The eggs are laid on the water hyacinth plants, typically near the plant tissue.
- They are small, oval-shaped, and are deposited in clusters.

2. Larval Stage

- Upon hatching, the larvae burrow into the plant tissues, feeding internally.
- The feeding activity creates tunnels within the plant, causing significant damage and reducing the plant's buoyancy.

3. Pupal Stage

- The larvae pupate either within the plant tissue or in the surrounding water, often creating a protective cocoon-like structure.
- This stage lasts until the larvae metamorphose into adults.

4. Adult Stage

- Adults emerge from the pupal stage fully formed and ready to continue the life cycle.
- They have a characteristic weevil appearance with a snout and are adapted to feed on the leaves of water hyacinth.

Expanded Flowchart for Mass Production of *Neochetina* Species

1. **Preparation of Growth Medium**
   - Fill outdoor tanks/plastic pools with water to a depth of 1 meter.
   - Add cow dung (200 g/m^3), superphosphate (40 g/m^3), and urea (10 g/m^3).
2. **Initial Exposure in Laboratory**
   - Place 25 water hyacinth plants in plastic troughs.
   - Release 50 *Neochetina* adults.
   - After one week, transfer adults to fresh plants.
3. **Multiplication Process**
   - Transfer exposed plants to outdoor tanks/pools.
   - Continue exposure until tanks/pools are filled.
   - Use adults for at least 15 exposures.

4. **Continuous Multiplication:**
   - Collect fresh adults every 3-4 months.
   - Maintain population with new plants as old ones dry.
   - Ensure continuous egg laying and subsequent generations.
5. **Collection and Field Release**
   - Collect 100-250 adults/month from each pool.
   - Release in fields to control water hyacinth.

Table: Developmental Stages and Observations

S.No	Stage	Duration	Observations
1	Egg Laying	5-10 days	Eggs laid on water hyacinth plants
2	Larval Feeding	2-3 weeks	Larvae feed internally, creating tunnels
3	Pupation	7-10 days	Pupae form within plant tissue or water
4	Adult Emergence	1-2 weeks	Adults emerge, ready to feed and mate
5	Adult Collection	Monthly	Collect adults for field release
6	Field Release	Continuous	Release adults to manage water hyacinth

Observation Record

Egg Collection

- Monitor and document the number of eggs laid daily.
- Record the frequency of leaf replacement and egg viability.

Plant Preparation

- Document the preparation process and health status of transplanted water hyacinth plants.
- Monitor the growth rate and health of the plants in the nutrient-enhanced water.

Larval Development

- Observe and record the feeding behavior and growth rate of larvae.
- Note the number of larvae per plant and their consumption rate, along with any damage to the plants.

Adult Emergence

- Track the emergence rate and number of adults collected daily.
- Record any anomalies or issues in the development process, such as deformities or parasitism.

Field Release

- Note the number of adults released and the frequency of releases.
- Monitor pest control effectiveness in the field, documenting reductions in water hyacinth coverage and plant health.

Field Collection and Preservation of Parasitoids and Predators

Methodology

1. Collection Areas

Conduct extensive field surveys for various insect pests of field crops and their insect parasitoids and predators in different districts of the Bundelkhand region. Target diverse ecosystems to gather a wide range of specimens.

2. Collection Equipment

The minimum equipment required for field collection includes:

- Pair of scissors
- Forceps
- Soft camel brush
- Collection tubes
- Polythene bags
- Paper bags
- Specimen jars
- Specimen tubes containing 75-90% alcohol (as a preservative)
- Muslin cloth
- Rubber bands
- Hand lens
- Absorbent paper
- Field diary
- Camera

3. Collection of Insect Predators of Crop Plants and Weed Insect Pests:

- **In Laboratory Rearing**
  - Some insect predators, such as aphid and whitefly predators, can be collected by rearing their immature stages along with the host pest in the laboratory.

- **In Field Collection**
  - Collect insect predators directly from the field while they are actively feeding on the host pest.

4. Collection of Insect Parasitoids of Crop Pests and Weeds

- Collect insect parasitoids from the host-insect pests through rearing processes in the laboratory. This involves maintaining the host-insect pests and allowing the parasitoids to emerge.

5. Cataloguing of Field and Laboratory Data

Maintain detailed records of breeding and field observations in catalogues. Use filing cards (6 cm x 4 cm) to document various information about the samples:

- Host-plant
- Host-pest
- Parasites and predators found
- Number of male and female parasites and predators
- Percentage of parasites and predators
- Location
- Altitude
- Remarks

Activities

List of Parasitoids and Predators Collected and Preserved

S.No	Name of Parasitoid/ Predator	Host Pest	Collected From
1	*Trichogramma chilonis*	Egg parasitoid of *Helicoverpa armigera*	Cotton fields
2	*Aphidius colemani*	Aphid parasitoid (*Aphis gossypii*)	Vegetable crops
3	*Chrysoperla carnea*	Predator of aphids, whiteflies, thrips	Cotton and vegetable fields
4	*Coccinella septempunctata*	Predator of aphids	Various crops
5	*Encarsia formosa*	Parasitoid of whiteflies	Greenhouses
6	*Bracon hebetor*	Larval parasitoid of *Helicoverpa armigera*	Maize and sorghum fields
7	*Orius insidiosus*	Predator of thrips	Flower crops
8	*Aphidoletes aphidimyza*	Predator of aphids	Fruit orchards
9	*Dicyphus hesperus*	Predator of whiteflies, thrips	Tomato fields

10	*Neoseiulus cucumeris*	Predator of spider mites	Greenhouses
11	*Phytoseiulus persimilis*	Predator of spider mites	Vegetable crops
12	*Cryptolaemus montrouzieri*	Predator of mealybugs	Citrus orchards
13	*Anagyrus lopezi*	Parasitoid of cassava mealybug	Cassava fields
14	*Eretmocerus eremicus*	Parasitoid of whiteflies	Greenhouses
15	*Amblyseius swirskii*	Predator of thrips and whiteflies	Greenhouses
16	*Harmonia axyridis*	Predator of aphids	Field crops
17	*Aenasius bambawalei*	Parasitoid of mealybugs	Cotton fields
18	*Bacillus thuringiensis*	Pathogen of caterpillars	Various crops
19	*Brachymeria lasus*	Parasitoid of *Spodoptera* spp.	Maize fields
20	*Metarhizium anisopliae*	Pathogen of various insect pests	Various crops

Field Collection and Preservation Process

1. Collection

- Use the appropriate collection equipment to gather specimens directly from the field. For predators, capture them while they are actively feeding on pests. For parasitoids, rear host pests in the laboratory until the parasitoids emerge.

2. Preservation

- Preserve collected specimens in specimen tubes containing 75-90% alcohol. Label each specimen tube with relevant information such as the date of collection, location, host pest, and any other pertinent details.

3. Cataloguing

- Maintain a detailed catalogue of all collected specimens. Use filing cards (6 cm x 4 cm) to record information such as host-plant, host-pest, number of male and female parasites and predators, percentage of parasites and predators, location, altitude, and any additional remarks.

4. Storage

- Store preserved specimens in a cool, dry place. Regularly check the specimens for any signs of degradation and replace the alcohol if necessary.

Illustrations of Different Life Stages of *Neochetina bruchi*

1. **Egg Stage**

- Small, oval-shaped eggs laid on the surface of water hyacinth leaves.

2. **Larval Stage**

- Grubs hatch from eggs and burrow into the plant tissues, feeding internally and creating tunnels.

3. **Pupal Stage**
   - Pupation occurs either within the plant tissue or in the surrounding water. Larvae form protective cocoons during this stage.

4. **Adult Stage**
   - Adults emerge from the pupal stage, ready to feed on water hyacinth leaves and continue the life cycle.

Mass Production of *Beauveria bassiana* (White Muscardine Fungus)

Materials Required

- Chalk powder
- Sorghum
- Water
- Autoclave
- Flasks (10-15)
- Hard cotton cushion

- Aluminum foil
- Nucleus culture of *Beauveria bassiana*

Methodology

1. Preparation of Sorghum Medium

1. **Soaking Sorghum:**
   - Soak 1 kg of sorghum in water for 48 hours.
   - Replace the water after the first 24 hours to ensure cleanliness and prevent microbial contamination.
2. **Draining Water**
   - After 48 hours, drain the water completely from the soaked sorghum.

2. Sterilization of Sorghum

3. **Separation and Sterilization**

- Divide the soaked sorghum equally into 10-15 flasks.
  - Plug each flask with a hard cotton cushion and wrap with double aluminum foil.
  - Sterilize the flasks for 40 minutes at 21 psi in an autoclave.

3. Inoculation and Growth

4. **Inoculation**
   - After sterilization, allow the flasks to cool.
   - Inoculate each flask containing sorghum with 2-3 drops of nucleus culture of *Beauveria bassiana*.
5. **Incubation**

- Allow the *Beauveria bassiana* culture to grow for 20-25 days. The fungal growth will appear as a white, fluffy mass over the sorghum grains.

4. Harvesting and Formulation

6. **Mixing with Chalk Powder**
   - Once the culture is fully grown, mix 2 kg of chalk powder with the *Beauveria* culture.
   - Ensure thorough mixing to create a homogeneous formulation.
7. **Drying**
   - Dry the mixture in the shade to preserve the viability of the fungal spores. Avoid direct sunlight to prevent UV damage to the spores.

Usage and Application

Dose

- For field application, use 1 gram of *Beauveria bassiana* formulation per liter of water or 1 kg per 1000 liters of water per hectare.
- Repeat the application at 10-20 day intervals for effective pest control.

Field Application

- **Preparation of Solution**
  - Dissolve the required amount of *Beauveria bassiana* formulation in water.
  - Stir well to ensure uniform distribution of the spores in the solution.
- **Spraying**
  - Use a sprayer to apply the solution uniformly over the target area.
  - Focus on areas where pest populations are high for maximum effectiveness.

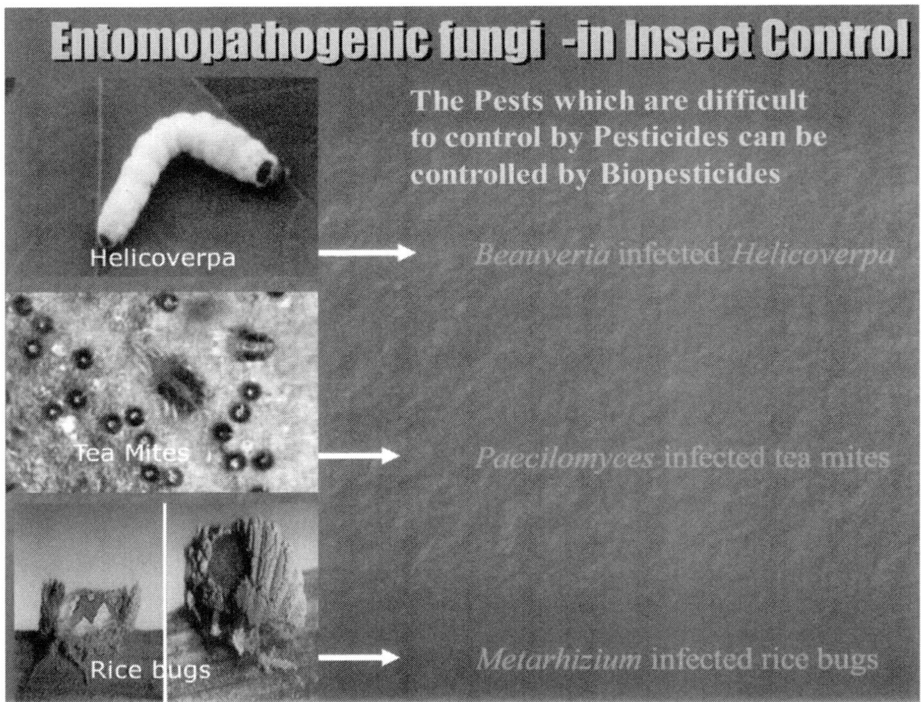

Diagram of an Insect Infected with *Beauveria bassiana*

- **Initial Infection**
  - The spores of *Beauveria bassiana* adhere to the insect cuticle.
  - The fungal spores germinate and penetrate the cuticle using enzymes and mechanical pressure.
- **Colonization**
  - The fungus grows inside the insect body, spreading through the hemolymph.
  - The insect's internal organs are gradually consumed by the fungal mycelium.
- **Sporulation**
  - The fungus emerges from the insect cadaver and produces spores.
  - The dead insect appears covered with white fungal growth, indicating successful infection and sporulation.

Mass Production of Nuclear Polyhedrosis Virus (Entomopathogenic Virus)

Virus Preparation Method

Insect Collection

Source of Insects

- **Laboratory Rearing:** The most efficient method involves rearing a large number of healthy *Helicoverpa armigera* larvae in laboratory conditions. However, this method is demanding and requires appropriate facilities.
- **Field Collection:** A cost-effective alternative is to collect larvae from pigeonpea or other crops through shaking operations.

Selection of Larvae

- For optimum HaNPV yield, select late fourth or early fifth instar larvae (about 1.0 – 1.5 cm long).

Inoculation of Virus

Preparation of Inoculum

- Soak chickpea seeds for 6-8 hours to prepare them for inoculation.
- Mix the seeds with a sufficient dose of virus inoculum (40 larval equivalent (LE)/kg of chickpea seed). Ensure thorough mixing to prevent underdosing or overdosing.

Infection Process

- Each insect is kept in an individual compartment with 2-3 virus-inoculated chickpea seeds.
- Label the compartments and leave the larvae to feed on the inoculated seeds and become infected.

Insect Rearing

Optimal Conditions

- Maintain rearing conditions at 25-35°C to promote virus production.
- Ensure that food is continuously available to minimize stress on the larvae.

Housing

- Rear infected larvae in multicellular trays with insect-proof coverings to prevent cannibalism and contamination.
- Use locally produced trays that can be stacked to save space.

Sanitation

- Clean the trays with disinfectant before and after rearing to prevent contamination by bacteria and protozoa.
- Implement fly-proofing measures to prevent fly oviposition on dead larvae.

Virus Harvesting

Collection of Infected Larvae

- Allow larvae to die naturally to ensure the virus completes its life cycle.
- Collect dead larvae showing clear symptoms of NPV, being careful not to damage them.
- Store collected larvae in refrigerated conditions to slow bacterial activity until further processing.

Processing Virus

Extraction

- Blend the collected dead larvae in a suitable blender to crush the insect tissue and release the NPV.
- Conduct the blending process in a cool room to maintain the virus's viability.
- Filter the blended material through a muslin cloth placed in a funnel to remove insect body parts. Squeeze the cloth gently to extract remaining liquid.

Centrifugation

- Centrifuge the filtrate at 3000-5000 rpm for 15 minutes to separate the virus from the homogenized liquid. This step helps remove bacteria and insect body fats that cause bad odors.
- Discard the turbid liquid at the top and save the paste-like substance at the bottom, which contains the NPV.

Storage

- Store the concentrated virus in UV-opaque bottles in refrigerated conditions. Properly label the bottles with relevant information.

Measurement of Quality of Virus

Quantification

- Use a haemocytometer and either a dark field or phase contrast microscope to count the polyhedral inclusion bodies (PIBs) per ml.
- In India, NPV is often quantified in terms of larval equivalents (LE). The proposed Indian standard for an LE is 6x10□ PIBs.

Bioassay

- Conduct bioassays to confirm the potency of the virus against target insects.

Application of NPV

Sensitivity to UV Light

- NPV is sensitive to ultraviolet rays. To improve effectiveness, apply NPV late in the day after peak sunshine.
- Adding UV absorbents such as 1 ml of robin blue to a liter of spray solution can also enhance effectiveness.

Dosage Recommendations

- For pigeonpea, use HaNPV at 3 x 10^{12} PIB (500 LE/ha).
- For chickpea, use HaNPV at 1.5 x 10^{12} PIB (250 LE/ha).

Summary Flowchart for Mass Production of NPV:

1. **Insect Collection**
   - Collect late fourth or early fifth instar larvae from fields or rear in the lab.
2. **Virus Inoculation**
   - Soak chickpea seeds for 6-8 hours.

- Mix seeds with virus inoculum (40 LE/kg).
- Feed larvae with inoculated seeds in individual compartments.

3. **Insect Rearing**
   - Maintain rearing conditions at 25-35°C.
   - Ensure constant food supply and use insect-proof trays.
4. **Virus Harvesting**
   - Allow larvae to die naturally.
   - Collect dead larvae showing NPV symptoms and refrigerate.
5. **Processing Virus**
   - Blend dead larvae and filter the mixture.
   - Centrifuge the filtrate to separate virus.
   - Store the virus in UV-opaque bottles in refrigerated conditions.
6. **Measurement and Application:**
   - Use a haemocytometer to count PIBs.
   - Conduct bioassays to confirm potency.
   - Apply NPV late in the day and add UV absorbents for improved effectiveness.

Table: Developmental Stages and Observations

S.No	Stage	Duration	Observations
1	Insect Collection	Ongoing	Collection of late instar larvae
2	Virus Inoculation	6-8 hours soaking	Mixing inoculum with chickpea seeds
3	Insect Rearing	25-35°C	Constant food supply, prevention of cannibalism
4	Virus Harvesting	Natural death	Collection and refrigeration of dead larvae
5	Processing Virus	15 minutes centrifuge	Blending, filtering, and centrifugation of larvae
6	Quality Measurement	Ongoing	Counting PIBs, conducting bioassays
7	Field Application	Late in the day	Applying NPV with UV absorbents

Observation Record

Insect Collection

- Record the source and number of larvae collected.
- Document the size and stage of the larvae.

Virus Inoculation

- Monitor the amount of inoculum used and its mixing efficiency with chickpea seeds.

Insect Rearing

- Track the rearing conditions (temperature, food availability) and record any issues with cannibalism or contamination.

Virus Harvesting

- Note the number of dead larvae collected and their condition.
- Document the refrigeration process.

Processing Virus

- Record the blending, filtering, and centrifugation procedures.
- Track the amount of virus concentrate obtained.

Quality Measurement

- Count PIBs using a haemocytometer.
- Conduct bioassays and record the results.

Field Application

- Note the timing of application and any UV protection measures used.
- Monitor the effectiveness of the NPV application in controlling target pests.

Quality Control and Registration Standards for Biological Control Agents

Importance of Quality Control

The use of biological control agents, such as insect parasitoids, predators, fungi, and viruses, is increasing worldwide. With the rise of mass production by various companies, ensuring the quality of these natural enemies is crucial. Quality control ensures that these agents function as intended after their release into the field. Poor quality can lead to failures in biological control programs, as seen with the failure of *Trichogramma brassicae* in Switzerland to control the European corn borer (*Ostrinia nubilalis*) in maize.

Quality Control Objectives

1. **Maintain Overall Quality**
   - Ensure that the overall quality of a species is maintained consistently.

2. **Identify Key Characteristics**
   - Determine the characteristics that affect the overall quality of the biological control agents.
3. **Assess Condition**
   - Evaluate whether the natural enemy is capable of controlling the pest to an acceptable level.

Development of Quality Control

Quality control in biological control can be approached from two sides:

1. **Functional Performance**
   - Measure how well the insect functions in its intended role. If it does not perform adequately, trace the cause and improve the rearing method.
2. **Expected Changes in Mass Rearing**
   - List the changes expected when mass rearing starts, measure these changes, and improve the rearing method if the changes are undesirable.

Registration Standards

Bio-Pesticides Registration

At present, microbial pesticides like fungi and NPV are included in the schedule of the Insecticides Act, 1968, and Insecticides Rule, 1971. These regulations ensure the quality of bio-pesticides at the farmer level. The Directorate of Plant Protection Quarantine and Storage (DPPQ&S), Department of Agriculture and Cooperation, Ministry of Agriculture, Government of India, has issued guidelines and data requirements for the registration of bio-pesticides in the country. All units must meet Indian standards and technical specifications to be eligible for registration under the Insecticides Act, 1968.

Registration of Insecticides

Application Process

1. **Form I Submission**
   - An application for registration of an insecticide must be made using Form I. The form, including the verification portion, must be signed by:
   - The individual applicant or their authorized representative.
   - The managing partner in the case of a partnership firm.

- A person authorized by the Board of Directors in the case of a company.
- The person in charge or responsible for the conduct of the business in other cases.

2. **Inspection of Testing Facility**
   - The Registration Committee may direct the inspection of the testing facility to establish the authenticity of the data.
3. **Application Fee**
   - Submit a duly filled application form with a bank draft of Rs. 100, drawn in favor of the Accounts Officer, DPPQ&S, payable at Faridabad. The application should be sent to the Secretary, Registration Committee, DPPQ&S, NH-IV, Faridabad-121001, Haryana. Include a self-addressed stamped envelope and one additional stamped envelope with the application.
4. **Registration Fee Payment**
   - Pay the registration fee by a demand draft drawn on the State Bank of India, Faridabad, in favor of the Accounts Officer, Directorate of Plant Protection, Quarantine and Storage, Faridabad, Haryana.
5. **Certificate of Registration**
   - The certificate of registration will be issued in Form II or Form II-A, as applicable, and will be subject to specified conditions.
6. **Duplicate Certificate**
   - A fee of Rs. 5 is required for a duplicate copy of the Certificate of Registration if the original is defaced, damaged, or lost.

Implementation of Quality Control Programs:

1. **Quality Control Checks**
   - Regularly monitor the quality of biological control agents through performance tests and inspections.
   - Ensure that the agents meet the established standards before release.
2. **Training and Awareness**
   - Train personnel involved in the mass production and application of biological control agents on quality control measures and the importance of maintaining high standards.
3. **Documentation and Record-Keeping**
   - Maintain detailed records of production processes, quality control checks, and field performance.

- Use this data to make informed decisions and improvements in rearing and application methods.

4. **Research and Development**
   - Continuously invest in research to improve the rearing techniques, formulation, and application methods of biological control agents.
   - Stay updated with the latest advancements and integrate them into the production process.

Examples of Quality Control Failures

1. ***Trichogramma brassicae* in Switzerland**
   - Failure to control the European corn borer (*Ostrinia nubilalis*) in maize due to poor quality control.
   - Highlighted the need for stringent quality control measures in the production and application of biological control agents.

Development of Insectaries and Their Maintenance

Insectaries are specialized facilities designed for the mass rearing of beneficial insects, such as parasitoids, predators, and pollinators, as well as insect pests for research and biological control purposes. Establishing and maintaining insectaries involves creating optimal environments that support the growth, reproduction, and health of these insects.

Part 1: Development of Insectaries

1.1 Planning and Design

1.1.1 Objectives and Scope

The first step in developing an insectary is to clearly define its objectives and scope. Understanding the primary purpose of the insectary will guide the design and operational procedures. Determine whether the facility will be used for rearing natural enemies, pollinators, insect pests for research, or a combination of these. This decision will influence the types of species to be reared, the scale of production, and the specific environmental conditions required.

Assessing the scope of operations involves estimating the number of species to be reared, the scale of production needed to meet research or commercial goals and the intended use of the insects. This assessment will help in designing the infrastructure and allocating resources effectively. For instance, rearing parasitoids for field release requires a different setup compared to maintaining insect pests for laboratory research. Additionally, consider the long-term goals of the insectary, such as potential expansion or diversification into rearing different species.

1.1.2 Location and Infrastructure

Selecting a suitable location is critical for the success of an insectary. The site should be easily accessible to researchers and staff, have a stable power supply, and be secured against contamination and unauthorized access. Proximity to fields or greenhouses where the insects will be used can also be beneficial.

The infrastructure of the insectary should include separate rooms or sections for different functions, such as quarantine, rearing, feeding, mating, and storage. Each room must be equipped with appropriate environmental controls to maintain optimal conditions for the insects. The design should facilitate efficient workflow and minimize the risk of cross-contamination between different insect populations. Additionally, consider the need for specialized areas, such as laboratories for research and development, or areas for preparing and storing diets and other supplies.

1.1.3 Environmental Control

Environmental control is paramount in an insectary. Install systems to regulate temperature, humidity, light, and ventilation, as these factors significantly impact insect growth and reproduction. Different insect species have specific environmental requirements; thus, the control systems should be programmable to maintain consistent conditions and allow for adjustments as needed.

For instance, many insect species thrive at temperatures between 25-30°C with relative humidity levels of 60-80%. Light conditions, including photoperiod and intensity, should be adjustable to simulate natural environmental cycles or specific experimental conditions. Ventilation systems should provide a constant supply of fresh air while preventing the entry of contaminants. Using high-efficiency particulate air (HEPA) filters can help maintain air quality and reduce the risk of airborne pathogens.

1.1.4 Biosecurity Measures

Implementing strict biosecurity measures is essential to prevent contamination by pathogens, parasites, and unwanted insects. Design entry and exit protocols, including airlocks, footbaths, and clothing changes, to minimize the risk of introducing contaminants. Each rearing room should have dedicated equipment to avoid cross-contamination, and staff should be trained in biosecurity protocols.

Regular monitoring and auditing of biosecurity measures are necessary to identify potential weaknesses and take corrective actions. In addition, establishing quarantine areas for new insect introductions allows for monitoring and ensuring they are free from diseases and parasites before integrating them into the main population. Biosecurity also involves managing waste effectively and ensuring that disposal methods prevent the spread of contaminants.

1.2 Facility Setup

1.2.1 Rearing Chambers

Rearing chambers or cages are essential components of an insectary. They should provide sufficient space and appropriate microenvironments for the insects being reared. The design and material of the chambers should facilitate easy cleaning and disinfection. Escape-proof features are crucial to prevent insects from escaping into the environment.

Different species may require different types of rearing chambers. For example, mesh cages are suitable for flying insects like butterflies and moths, while solid-walled chambers may be used for crawling insects. Ensure that the chambers are equipped with feeders and water sources appropriate for the insect species. For species that require specific substrates for egg-laying or pupation, provide suitable materials such as sand, soil, or plant leaves.

1.2.2 Equipment and Supplies

Equip the insectary with all necessary tools and supplies, such as rearing containers, feeding equipment, sterilization tools, and monitoring devices. Maintaining an inventory of consumables, including diets, growth media, and disinfectants, ensures continuous operations without interruptions.

Specialized equipment, such as microscopes for inspecting insect health and development, climate control devices, and automated feeders, can enhance the efficiency and effectiveness of rearing operations. Additionally, safety equipment such as gloves, lab coats, and eye protection should be readily available to protect staff and prevent contamination.

1.2.3 Quarantine and Isolation

Quarantine and isolation areas are vital for managing new insect introductions and handling potential outbreaks of disease or contamination. New insects should be quarantined for a period to monitor for any signs of illness or parasitism before being introduced to the main population.

Isolation protocols should be established for infected or infested insects to prevent the spread of contaminants within the insectary. This includes using dedicated equipment and separate rearing chambers for isolated insects. Regular health checks and rapid response to any issues can help maintain the overall health and productivity of the insectary.

1.2.4 Environmental Chambers

For species requiring specific photoperiods or temperature regimes, environmental chambers can simulate seasonal variations and other ecological conditions. Equip these chambers with programmable lights and temperature controls to create tailored rearing conditions.

Environmental chambers can replicate the natural environment of the insects, facilitating research on life cycle stages, behavior, and interactions with other organisms. They can also be used to test the resilience of insects to different environmental stressors, such as temperature fluctuations or changes in humidity.

1.2.5 Ventilation Systems

Proper ventilation is crucial to maintaining air quality and preventing the buildup of harmful gases such as carbon dioxide and ammonia. Install high-efficiency particulate air (HEPA) filters to remove airborne contaminants and maintain a clean environment.

Ventilation systems should provide a constant supply of fresh air while maintaining the desired temperature and humidity levels. Regular maintenance and monitoring of the ventilation system are necessary to ensure its effectiveness and prevent mechanical failures that could compromise the rearing conditions.

1.2.6 Water Management

Many insect species require humidity control, which may involve using humidifiers or dehumidifiers. Ensure a reliable water supply for cleaning, humidification, and any aquatic insect rearing requirements.

Water management also involves preventing waterlogging and mold growth in rearing chambers. Proper drainage systems should be in place to remove excess water, and regular checks should be conducted to ensure that humidity levels are within the optimal range for the reared species.

Part 2: Maintenance of Insectaries

2.1 Routine Operations

2.1.1 Environmental Monitoring

Regularly monitoring environmental conditions is critical for the successful rearing of insects. Use data loggers and environmental sensors to track temperature, humidity, light, and ventilation continuously. Consistent monitoring allows for immediate adjustments if conditions deviate from the optimal range.

Analyze the collected data to identify trends and potential issues. For example, fluctuations in temperature might indicate problems with the HVAC system, while changes in humidity could be linked to water management issues. Addressing these problems promptly can prevent adverse effects on insect health and productivity.

2.1.2 Feeding and Nutrition

Providing a balanced diet tailored to the nutritional needs of each insect species is essential for their growth and reproduction. Diets may include artificial diets, plant material, or prey insects. Regularly schedule feeding times and monitor the health and growth of the insects to ensure they are receiving adequate nutrition.

Developing and maintaining consistent feeding protocols can improve the overall health and performance of the insects. Document the feeding schedules, diet composition, and any observed changes in insect behavior or health. Adjust the diet composition as needed based on these observations.

2.1.3 Sanitation and Hygiene

Implement a rigorous cleaning schedule to maintain a sterile environment. Clean and disinfect rearing chambers, equipment, and work surfaces regularly. Dispose of waste material, such as frass and uneaten food, promptly to prevent the buildup of contaminants.

Sanitation protocols should include specific steps for cleaning different types of equipment and surfaces. Use appropriate disinfectants that are effective against common pathogens but safe for the insects being reared. Regularly review and update the sanitation protocols to incorporate new knowledge and technologies.

2.1.4 Health Monitoring

Regularly inspect insect populations for signs of disease, parasitism, and abnormal behavior. Isolate and treat affected individuals promptly to prevent the spread of infections. Maintain records of health issues and treatments to identify patterns and prevent future outbreaks.

Health monitoring should involve routine checks for physical abnormalities, behavioral changes, and signs of illness. Use diagnostic tools, such as microscopes and molecular tests, to accurately identify pathogens and parasites. Implement preventive measures, such as vaccinations or probiotics, if applicable.

2.1.5 Pest Control

Implement integrated pest management (IPM) strategies within the insectary to control unwanted pests without harming the reared insects. Use physical barriers, traps, and biological control agents as primary methods. Chemical controls should be a last resort and used with caution.

Regularly inspect the insectary for signs of pest infestations, such as droppings, damaged plants, or unusual insect activity. Develop a pest management plan that includes monitoring, prevention, and control measures. Train staff on identifying and responding to pest issues promptly and effectively.

2.2 Record-Keeping and Documentation

2.2.1 Data Management

Keep detailed records of environmental conditions, feeding schedules, health status, and breeding outcomes for each insect species. Use electronic databases or logbooks to organize and store data, making it easily accessible for analysis and reporting.

Implement a standardized data recording system to ensure consistency and accuracy. Regularly back up electronic data to prevent loss due to technical issues. Analyze the data periodically to identify trends and make informed decisions about rearing practices and improvements.

2.2.2 Breeding and Production Records

Document the breeding history and production rates of each insect species. Record the number of eggs laid, larvae hatched, pupae formed, and adults emerged. Track the performance of different breeding pairs or colonies to identify successful and unsuccessful lines.

Maintain separate records for each insect species and strain to facilitate comparison and analysis. Use this data to optimize breeding strategies and improve the overall productivity and quality of the reared insects. Share the findings with relevant stakeholders to enhance collaborative efforts.

2.2.3 Quality Control

Implement quality control measures to ensure that reared insects meet the desired standards. Conduct regular assessments of insect size, vigor, reproductive capacity, and pest control efficacy. Use control groups and standard protocols to evaluate the performance of insects before they are released or used in experiments.

Develop and follow standard operating procedures (SOPs) for quality control assessments. Train staff on the importance of quality control and how to conduct assessments accurately. Regularly review and update the SOPs based on new research and feedback from quality control assessments.

2.2.4 Performance Metrics

Develop key performance indicators (KPIs) to evaluate the success of the rearing program. These might include survival rates, reproductive success, growth rates, and efficacy in pest control applications. Regularly review and analyze performance data to identify trends and areas for improvement.

Set benchmarks for each KPI based on historical data and industry standards. Use the KPIs to track progress over time and make data-driven decisions about rearing practices. Share performance metrics with stakeholders to demonstrate the effectiveness and impact of the rearing program.

2.3 Maintenance of Biosecurity

2.3.1 Contamination Prevention

Continue to enforce biosecurity measures rigorously. Ensure that all personnel follow entry and exit protocols and that all equipment is cleaned and sterilized before and after use. Conduct regular audits to identify potential biosecurity breaches and take corrective actions.

Biosecurity audits should include inspections of rearing chambers, equipment, and storage areas. Use checklists to ensure that all biosecurity measures are followed consistently. Address any identified issues promptly to prevent the spread of contaminants and maintain a healthy rearing environment.

2.3.2 Emergency Response

Develop and implement an emergency response plan for dealing with outbreaks of diseases, pests, or contamination. Train staff on emergency procedures and ensure that necessary supplies are readily available. Isolate affected areas immediately and follow containment protocols to prevent the spread of contaminants.

The emergency response plan should include clear steps for identifying, containing, and eliminating threats. Designate a response team responsible for managing emergencies and communicating with staff and stakeholders. Conduct regular drills to test the effectiveness of the response plan and make improvements as needed.

2.3.3 Biosecurity Drills

Conduct regular biosecurity drills to prepare staff for potential contamination events. These drills should simulate real-world scenarios and test the effectiveness of response protocols. Review and refine biosecurity plans based on drill outcomes to address any weaknesses.

Biosecurity drills should involve all staff and cover different types of contamination scenarios. Use the outcomes of the drills to identify gaps in training, protocols, and resources. Continuously improve the biosecurity plans to enhance the overall resilience of the insectary.

2.3.4 Collaborative Research

Collaborate with other insectaries and research institutions to share best practices, innovations, and research findings. This collaboration can help improve biosecurity measures and rearing techniques, contributing to the overall success of biological control programs.

Engage in joint research projects, conferences, and workshops to stay updated on the latest developments in insect rearing and biosecurity. Share data and experiences with peers to foster a culture of continuous improvement and innovation. Establish formal partnerships to leverage resources and expertise.

Multiple Choice Questions (MCQs)

1. What is the primary objective of developing insectaries?

   A) To grow food crops

   B) To mass-produce beneficial insects

   C) To conduct human medical research

   D) To study plant genetics

   Answer: B

2. Which of the following is a common material used for insect rearing chambers?

   A) Metal sheets B) Porous wood

   C) Glass D) Non-porous plastic

   Answer: D

3. What is a critical environmental factor to control in insectaries for optimal insect growth?

   A) Soil pH B) Light intensity

   C) Temperature D) Wind speed

   Answer: C

4. Why is it important to implement biosecurity measures in insectaries?

   A) To reduce the cost of rearing insects

   B) To prevent contamination and spread of diseases

   C) To increase the size of insect populations

   D) To improve the aesthetic appearance of the facility

   Answer: B

5. In the mass production of *Beauveria bassiana*, what is the purpose of mixing the culture with chalk powder?

   A) To enhance the color of the fungus

   B) To improve the storage and application properties

   C) To increase the nutritional value for insects

   D) To neutralize the pH of the culture

   Answer: B

6. Which equipment is essential for monitoring the quality of *Beauveria bassiana* spores?

   A) Spectrophotometer B) Haemocytometer

   C) Autoclave D) pH meter

   Answer: B

7. What is the main goal of quality control programs for bio-control agents?

   A) To reduce production costs

   B) To ensure agents perform effectively in the field

   C) To increase the diversity of agents produced

   D) To enhance the appearance of agents

   Answer: B

8. How often should environmental conditions in insectaries be monitored?

   A) Monthly B) Annually

   C) Daily D) Bi-weekly

   Answer: C

9. Which of the following is NOT a method for field release of bio-control agents?

   A) Directly spraying the agents onto crops

   B) Releasing agents from aircraft

   C) Introducing predators into confined cages in the field

   D) Broadcasting through irrigation systems

   Answer: C

10. What is the primary benefit of using biological control agents in pest management?

    A) They are cheaper than chemical pesticides

    B) They enhance the growth rate of crops

    C) They provide a sustainable and eco-friendly solution

    D) They eliminate the need for any other pest control methods

    Answer: C

11. During the mass production of *Trichogramma* species, why are eggs treated under UV light?

    A) To sterilize the eggs

    B) To enhance hatching rates

    C) To kill pathogens

    D) To induce genetic mutations

    Answer: A

12. What is the significance of using control groups in quality control of bio-control agents?

    A) To reduce costs

    B) To provide a benchmark for performance evaluation

    C) To increase production speed

    D) To diversify the agents

    Answer: B

13. Which factor is critical for maintaining the health of insects in rearing facilities?

    A) Regular exposure to sunlight

    B) Consistent availability of food

    C) High population density

    D) Use of natural predators

    Answer: B

14. Why is it important to store biological control agents in UV-opaque bottles?

    A) To prevent overheating

    B) To protect from UV light degradation

    C) To increase shelf life

    D) To reduce costs

    Answer: B

15. What is a common challenge in mass rearing bio-control agents?

    A) Lack of suitable rearing materials

    B) Maintaining genetic diversity

    C) Ensuring consistent environmental conditions

    D) Both B and C

    Answer: D

16. What is a significant economic consideration in the mass production of bio-control agents?

    A) Cost of initial setup

    B) Market demand for bio-control agents

    C) Operational costs

    D) All of the above

    Answer: D

17. How is the effectiveness of a bio-control agent typically evaluated in the field?

    A) By the speed of agent release

    B) By the reduction in pest population

    C) By the aesthetic improvement of crops

    D) By the rate of agent reproduction

    Answer: B

18. Which insect is commonly used for producing Nuclear Polyhedrosis Virus (NPV)?

    A) *Helicoverpa armigera* B) *Aphis gossypii*

    C) *Drosophila melanogaster* D) *Tetranychus urticae*

    Answer: A

19. What is the primary purpose of using HEPA filters in insectaries?

    A) To increase humidity

    B) To improve lighting conditions

    C) To remove airborne contaminants

    D) To regulate temperature

    Answer: C

20. In the context of insect rearing, what does the term "quarantine" refer to?

    A) A method to speed up insect growth

    B) A technique for genetic modification

    C) Isolation of new insects to monitor for diseases

    D) A type of diet for insects

    Answer: C

Short Questions

1. What are the primary objectives of developing an insectary?
2. Name two key environmental factors that must be controlled in an insectary.
3. Why is it important to implement biosecurity measures in insectaries?
4. What is the purpose of using a haemocytometer in the production of *Beauveria bassiana*?
5. Which insect is commonly used for producing Nuclear Polyhedrosis Virus (NPV)?

Long Questions

1. Describe the process of mass producing *Beauveria bassiana*, including the steps involved in preparing the inoculum, rearing the fungus, and formulating the final product for application. Discuss the importance of each step in ensuring the quality and effectiveness of the bio-control agent.
2. Explain the key components and considerations involved in developing an insectary. Include details on planning and design, environmental control, biosecurity measures, and the necessary equipment and supplies. Discuss how each component contributes to the successful rearing of high-quality insects for biological control programs.
3. Discuss the importance of quality control in the mass production of bio-control agents. What are the main goals of quality control programs, and what methodologies are used to ensure that bio-control agents perform effectively in the field? Provide examples of potential failures in biological control due to poor quality control and how these can be prevented through rigorous quality control measures.

Unit IV

Successful biological control projects, analysis, trends and future possibilities of biological control. Importation of natural enemies- Quarantine regulations, biotechnology in biological control. Semiochemicals in biological control.

A. Successful Biological Control Projects

Biological control, the use of natural enemies to manage pest populations, has seen numerous successful implementations globally for example:

1. Control of Cottony Cushion Scale (*Icerya purchasi*)

The case of the cottony cushion scale, *Icerya purchasi*, in California's citrus orchards serves as a seminal example of biological control's efficacy. Originating from Australia, this pest posed a severe threat to the burgeoning citrus industry in California in the late 19th century. The story of its control through the introduction of natural enemies is not only a landmark in biological control history but also a demonstration of ecological problem-solving.

Background and Impact

Icerya purchasi is a sap-sucking insect that infests a wide range of host plants but shows a particular fondness for citrus trees. The infestation is damaging as the scale excretes a sticky substance known as honeydew, which promotes the growth of sooty mold, impairing photosynthesis and weakening the plants. The severity of the infestation threatened the economic viability of the citrus industry in California during the 1880s.

Introduction of Natural Enemies

The turning point came when Albert Koebele, a biologist working for the United States Department of Agriculture, was sent to Australia in search of natural predators that could control the scale. In 1888, two key biological control agents were identified and subsequently introduced to California:

1. **The Vedalia Beetle (*Rodolia cardinalis*)**
   - This beetle, particularly its larvae, proved highly effective in controlling *Icerya purchasi*. The larvae voraciously consume the scale insects at all stages of their lifecycle, from eggs to adults. The

introduction of the Vedalia beetle led to dramatic reductions in the scale population. The lifecycle of the beetle is well-synced with that of the scale, ensuring sustained control over the pest population.

2. **The Parasitic Fly (*Cryptochetum iceryae*)**
   - Although less celebrated than the Vedalia beetle, this parasitic fly also plays a crucial role in suppressing the cottony cushion scale populations. The fly lays its eggs on the scale, and the emerging larvae consume the scale from within. This parasitization contributes significantly to the overall mortality rates of the scale population.

Outcomes and Legacy

The introduction of these biological control agents led to rapid and substantial reductions in the *Icerya purchasi* population. By the early 1890s, the scale was under complete control, and the citrus industry was saved from probable devastation. This success story was one of the first to be documented and helped to establish biological control as a viable and environmentally friendly strategy for pest management.

The control of *Icerya purchasi* not only saved the California citrus industry but also set a precedent for the use of biological control worldwide. It demonstrated that carefully selected and introduced natural enemies could effectively manage pest populations without the need for harmful chemical pesticides, thereby preserving the ecological balance and promoting sustainable agricultural practices.

2. Control of *Phenacoccus manihoti* in African Cassava Crops

The control of the cassava mealybug, *Phenacoccus manihoti*, in Africa is a prominent example of biological control addressing a severe agricultural pest problem. Originally from South America, the mealybug was inadvertently introduced into Africa, where it rapidly became a major pest, devastating cassava crops—a critical food source for millions.

Background and Impact

Cassava, a staple food crop in many African countries, is known for its resilience to harsh weather conditions and poor soils, making it an essential sustenance crop. However, the introduction of the *Phenacoccus manihoti* disrupted cassava production due to the mealybug's destructive feeding habits, which stunted growth and led to significant yield losses. The infestation quickly reached a critical level, threatening food security across the continent.

Introduction of *Apoanagyrus lopezi*

In response to this crisis, researchers and agricultural experts turned to biological control as a sustainable solution. A key figure in this effort was Hans Herren and his team at the International Institute of Tropical Agriculture, who identified the parasitoid wasp *Apoanagyrus lopezi* (formerly known as *Epidinocarsis lopezi*) as a potential biological control agent. This wasp was native to South America, the original home of the cassava mealybug.

1. **Selection and Importation**:
   - Extensive research was conducted to ensure that *Apoanagyrus lopezi* would specifically target the cassava mealybug without negatively impacting other species. After promising results in controlled trials, the wasp was introduced to Africa.
2. **Mechanism of Action**
   - The female wasp lays eggs inside the mealybug. When the eggs hatch, the wasp larvae consume the mealybug from the inside, eventually killing the host. This biological process significantly reduces mealybug populations without the need for chemical pesticides.

Outcomes and Legacy

The introduction of *Apoanagyrus lopezi* proved highly effective. Within a few years of its release, there was a dramatic decline in mealybug populations, leading to a substantial recovery of cassava yields. This success story not only safeguarded the food security of millions but also demonstrated the potential of biological control in managing invasive pests sustainably.

The recovery of cassava crops following the introduction of *Apoanagyrus lopezi* highlighted several key aspects:

- **Sustainability**: The approach provided a long-term solution that reduced dependency on chemical pesticides, which can be harmful to the environment and human health.
- **Economic Impact**: Restoring cassava yields helped stabilize the livelihoods of countless farmers and contributed to the economic stability of affected regions.
- **Scalability and Replicability**: The success set a precedent for other regions facing similar invasive pest challenges.

3. Management of *Hypericum perforatum* in U.S. Rangelands

The successful management of Klamath weed, scientifically named *Hypericum perforatum* and also known as St. John's wort, stands as a classic example of effective biological control in controlling invasive plant species. This toxic plant, which had extensively invaded U.S. pastures and rangelands, posed significant challenges due to its toxicity to livestock and displacement of native flora.

Problem Background

Originally from Europe, *Hypericum perforatum* adapted well to the U.S. environment, lacking natural competitors or predators. Its presence in pastures led to reduced grazing areas, as the plant's toxicity deterred livestock feeding and could lead to poisoning if ingested. This weed's rapid spread and dense coverage necessitated urgent control measures to protect the economy and the ecology of affected areas.

Biological Control Strategy

In response to this burgeoning issue, entomologists and biologists turned to the weed's native habitat to identify natural predators. Two beetle species, both known to feed on Hypericum perforatum specifically, were selected for introduction to the U.S.:

1. ***Chrysolina quadrigemina***:
   - Known as the St. John's wort beetle, this beetle consumes the leaves and flowers of the weed. Its feeding activity is synchronized with the weed's growth cycles, making it an effective biological control agent.
2. ***Chrysolina hyperici***
   - This beetle also feeds on the foliage of Klamath weed and adapts well across different environmental conditions. This adaptability ensures that the beetle is effective in various regions, offering a widespread reduction in weed prevalence.

Control Mechanism

The control mechanism involves the beetles consuming the plant's foliage, impairing its ability to photosynthesize and reproduce. The larvae of these beetles are particularly effective, consuming the plant tissues extensively. This targeted feeding significantly reduces the vitality of the weed, limiting both its growth and reproductive capabilities.

Results and Impact

The deployment of *Chrysolina quadrigemina* and *Chrysolina hyperici* led to a substantial decrease in the abundance of *Hypericum perforatum* across infested regions. The decline in weed populations allowed native vegetation to recover, restoring the ecological balance and usable land for grazing. This recovery not only benefited the environment but also supported local economies by increasing the availability of safe, productive grazing land.

B. Analysis of Successful Biological Control Projects

Analyzing successful biological control projects reveals several key factors that contribute to their success:

1. Selection of Effective Natural Enemies

Choosing the right natural enemies for biological control is fundamental to successful pest management strategies. The effectiveness of a biological control agent hinges primarily on two key characteristics: host specificity and the ability to effectively locate and suppress pest populations.

Host Specificity

Host specificity is paramount in the selection of biological control agents. This characteristic ensures that the chosen agent targets only the intended pest species without affecting non-target species, which could include beneficial insects or plants. High host specificity is crucial to avoid unintended ecological consequences, such as the decline of non-target species or disruptions to the local biodiversity. Agents with high specificity are less likely to cause ecological imbalances and are more acceptable from an environmental safety perspective.

Search and Attack Efficiency

The ability of biological control agents to efficiently search for and attack pest populations is another critical factor. This capability ensures that the agents can effectively reduce pest numbers and achieve control over the infestation. Factors that enhance an agent's search efficiency include:

- **Mobility**: The ability to move across varying landscapes and conditions to find the pest.
- **Perceptual abilities**: Strong sensory capabilities to detect host signals (such as pheromones or visual cues).
- **Reproductive rate**: The potential to produce sufficient offspring to maintain pressure on the pest population.
- **Survival skills**: Adaptability to the local environment, including climatic conditions and availability of resources.

These characteristics determine how well a biological control agent can sustain its population in a new environment and achieve the desired control level over the pest.

Implications for Biological Control Programs

The careful selection of natural enemies based on these criteria is crucial for the development of effective and sustainable biological control programs. By prioritizing host specificity and search-and-attack efficiency, practitioners can design interventions that are both effective in controlling pest populations and minimizing risks to non-target species and the environment.

2. Comprehensive Research and Testing in Biological Control

Before introducing a biological control agent into a new environment, comprehensive research, and rigorous testing are indispensable to ensure both the efficacy and safety of the intervention. This preparatory phase encompasses a thorough understanding of the biology and ecology of the pest and the control agent, alongside extensive host specificity testing and environmental impact assessments.

Understanding Biology and Ecology

A deep understanding of the biological and ecological characteristics of both the pest and the potential control agent forms the foundation of any successful biological control program. Key aspects include:

- **Life cycle**: Knowledge of the life cycle stages of the pest and the control agent helps in timing the introduction and management of the biological control measures.
- **Behavioral patterns**: Insights into feeding, mating, and nesting behaviors facilitate the prediction and manipulation of interactions between the pest and the control agent.
- **Ecological interactions**: Understanding the interactions with other species and the ecosystem at large ensures that the control agent's introduction will not disrupt existing balances.

Host Specificity Testing

Host specificity testing is crucial to ascertain that the control agent targets only the intended pest species, minimizing the risk to non-target organisms. This testing involves:

- **Laboratory trials**: Initial screening in controlled environments to observe the behavior of the control agent with potential host and non-host species.
- **Field trials**: Testing in a more natural setting to verify the laboratory findings and observe longer-term impacts and interactions.

Environmental Impact Evaluation

Evaluating the potential environmental impacts of introducing a new biological control agent is critical to prevent any adverse effects on the ecosystem. Considerations include:

- **Potential for invasiveness**: Assessing whether the control agent might become invasive, affecting native species and habitats.
- **Impact on non-target species**: Studying possible unintended effects on beneficial insects, native plants, or animals.
- **Long-term ecological effects**: Understanding the broader ecological implications, including potential shifts in habitat or species dynamics.

Implementation Considerations

Only after thorough research and testing should a biological control agent be released. This phased approach ensures that the agent will be both safe and effective, supporting sustainable pest management without unintended consequences. Regulatory approvals, guided by scientific data from these studies, are also crucial to align with both national and international environmental protection standards.

3. Integration with Other Control Methods in Biological Control

Integrating biological control efforts with other pest management strategies forms the core of an Integrated Pest Management (IPM) approach. This method enhances the overall effectiveness and sustainability of pest control by combining multiple strategies that work synergistically. Successful biological control projects often incorporate cultural practices, chemical controls, and habitat manipulation to achieve comprehensive and enduring pest suppression.

Cultural Practices

Cultural practices involve modifying farming techniques or habitat conditions to make the environment less hospitable to pests. Examples include:

- **Crop rotation**: Changing the types of crops grown in successive seasons to disrupt pest life cycles.
- **Sanitation**: Removing plant debris or residues that may harbor pests or provide them with breeding sites.
- **Planting pest-resistant crop varieties**: Using genetically or traditionally bred crops that are less susceptible to pest attacks.

These practices can reduce pest populations and their impact, making biological control more effective by limiting the opportunities for pest resurgence.

Chemical Controls

While biological control aims to reduce reliance on chemical pesticides, judicious use of chemicals can still play a role in an IPM approach. The key is to use them in a way that does not harm the biological control agents. Strategies include:

- **Targeted applications**: Using pesticides in a localized manner or during times when control agents are least active or vulnerable.
- **Using selective pesticides**: Choosing chemicals that are less toxic to the natural enemies introduced for biological control.
- **Reduced dosage and frequency**: Minimizing the use of chemicals to the necessary levels to manage pests without disrupting biological control processes.

Chemical controls, when used sparingly and strategically, can support biological methods by quickly reducing pest populations to manageable levels.

Habitat Manipulation

Habitat manipulation involves altering the physical environment to enhance the effectiveness of biological control agents. Techniques include:

- **Creating refuges for natural enemies**: Planting cover crops or maintaining hedgerows can provide shelter and alternative food sources for predatory insects and parasitoids.
- **Enhancing biodiversity**: Incorporating a variety of plant species can attract and sustain a broader range of natural enemies, thereby improving pest control.
- **Water management**: Adjusting irrigation practices to favor natural enemies or disadvantage pests.

These modifications can improve the survival and efficacy of biological control agents, making the pest management system more robust and resilient.

Benefits of an Integrated Approach

An IPM approach that includes biological control along with other methods offers several advantages:

- **Reduced pesticide use**: Integration reduces reliance on chemical pesticides, lowering environmental toxicity and the risk of pesticide resistance.
- **Long-term sustainability**: By using a combination of strategies, IPM promotes ecological balance and reduces the likelihood of pest populations rebounding.

- **Cost-effectiveness**: Over time, the reduced need for inputs like pesticides can lower the cost of pest management.

4. Continuous Monitoring and Management in Biological Control

Ongoing monitoring and management are critical components of a successful biological control program. This proactive approach ensures that the effectiveness of the biological control agents is maintained over time and allows for timely adjustments in response to changing conditions or unforeseen challenges. This process involves several key activities, including tracking populations of both pests and control agents, evaluating environmental impacts, and addressing emerging issues.

Tracking Pest and Control Agent Populations

Regular monitoring of both pest and control agent populations is essential to assess the success of the biological control efforts and to detect any signs of pest resurgence or declines in control agent effectiveness. This can involve:

- **Population surveys**: Regular field inspections and data collection to measure population dynamics of both the pest and the biological control agents.
- **Use of monitoring tools**: Deployment of traps, visual surveys, and other tools to quantitatively assess population levels.
- **Data analysis**: Analyzing collected data to identify trends, effectiveness, and any need for intervention.

These activities help to ensure that the control agents are effectively suppressing the pest population and provide early warning signs of any issues that may require intervention.

Evaluating Environmental Impacts

As part of the ongoing management, it is crucial to continually assess the environmental impacts of the biological control program. This includes:

- **Impact on non-target species**: Monitoring ecosystems to ensure that non-target organisms are not adversely affected by the introduced control agents.
- **Biodiversity assessments**: Evaluating the effects of the biological control on local biodiversity, ensuring that the introduction of control agents does not disrupt ecological balances.
- **Long-term ecological monitoring**: Establishing long-term monitoring programs to observe the broader ecological impacts and ensure sustainable outcomes.

Environmental monitoring helps maintain ecological integrity and public trust in biological control practices.

Addressing Emerging Issues

Adaptive management is a key feature of successful biological control programs. This involves:

- **Responding to resistance**: If pests develop resistance to the control agents, it may be necessary to introduce additional or alternative agents.
- **Adjusting control strategies**: Based on monitoring data, it may be necessary to modify the deployment of control agents, enhance their habitat, or revise management practices to improve effectiveness.
- **Stakeholder engagement**: Keeping communication open with farmers, land managers, and the public to address concerns and provide updates on the status and effectiveness of the control measures.

Importance of Continuous Monitoring

Continuous monitoring and adaptive management are vital because biological control is not a "set and forget" solution. It requires active oversight to adapt to any changes in the pest populations, environmental conditions, or other factors that could impact the effectiveness of the control agents. This vigilance helps to ensure that the biological control measures remain effective over time, preventing the resurgence of pests and minimizing any negative ecological impacts.

C. Trends in Biological Control

Several trends are shaping the future of biological control:

1. Advancements in Molecular Biology

Molecular biology is significantly influencing the field of biological control, offering new tools and methodologies that enhance the precision and effectiveness of pest management strategies. Several key advancements in this area are reshaping how biological control agents are selected, monitored, and utilized.

DNA Barcoding

DNA barcoding is a technique used to identify species based on a short, standardized region of their DNA. In the context of biological control, DNA barcoding enables:

- **Accurate Identification**: Quick and precise identification of both pests and potential control agents, crucial for ensuring the specificity of biological control measures.

- **Detection of Cryptic Species**: Identification of species that appear identical to the naked eye but are genetically distinct. This is especially important in selecting the most effective control agents for specific pest populations.

DNA barcoding enhances the accuracy of biological control programs by ensuring that the correct species are targeted and monitored, thereby improving the overall efficacy of pest management efforts.

Genomic Analysis

Genomic analysis involves examining the complete genetic makeup of organisms, providing deep insights into their biology, behavior, and interactions with the environment. In biological control, genomic analysis helps by:

- **Understanding Pest and Agent Interactions**: Elucidating the genetic basis of host specificity and resistance mechanisms, which can lead to the development of more effective and durable control strategies.
- **Enhancing Agent Selection**: Identifying genetic traits that predict a potential control agent's adaptability, survival, and effectiveness in various environmental conditions.

CRISPR and Genetic Engineering

The advent of CRISPR and other genetic engineering tools has opened up possibilities for directly manipulating the genetics of biological control agents. Potential applications include:

- **Enhancing Efficacy**: Engineering control agents to enhance their resilience against environmental stresses or improve their efficacy against pests.
- **Targeted Pest Control**: Developing genetically modified organisms that can target pests in a highly specific manner, potentially reducing the impact on non-target species and increasing the safety of biological control programs.

Implications for the Future

The integration of molecular biology techniques into biological control strategies represents a significant advancement in the field. These techniques not only improve the selection and monitoring of control agents but also hold the promise of creating novel agents with enhanced capabilities. As molecular tools continue to evolve, they will likely play a central role in developing safer, more effective, and more sustainable biological control solutions.

2. Increased Focus on Conservation Biological Control

Conservation biological control is gaining momentum as a key trend in the field, emphasizing the enhancement of natural enemies' effectiveness through environmental modifications. This approach aligns with broader environmental and sustainability goals, focusing on creating ecosystems that naturally regulate pest populations. The trend towards conservation biological control involves several key practices, each contributing to a more sustainable and effective pest management strategy.

Planting Cover Crops

Cover crops play a pivotal role in conservation biological control by providing several ecological benefits:

- **Enhanced Biodiversity**: Cover crops contribute to increased plant diversity, which can attract and support a wider range of natural enemies.
- **Alternative Food Sources**: They provide additional nectar, pollen, and habitat for predatory and parasitic insects, extending their longevity and efficacy in controlling pests.
- **Improved Soil Health**: Healthy soils support more robust plant growth and can better support diverse ecosystems, including beneficial insects.

Creating Habitats for Natural Enemies

Developing specific habitats designed to support natural enemies is another crucial aspect of conservation biological control. This can include:

- **Hedgerows and Floral Strips**: Planting rows of shrubs or flowers alongside or within crop fields to provide shelter, breeding sites, and alternative food sources for natural enemies.
- **Woodlots and Water Features**: Maintaining or introducing features like ponds or small woodlands near agricultural areas to support a broader range of biodiversity, contributing to the stability and resilience of natural enemy populations.

Reducing Pesticide Use

A central component of conservation biological control is the reduction or strategic use of chemical pesticides to minimize their impact on non-target species, including natural enemies. This can be achieved through:

- **Targeted Pesticide Application**: Using pesticides only when absolutely necessary and in targeted areas to avoid widespread effects on beneficial organisms.

- **Selecting Safer Pesticides**: Opting for biopesticides or chemicals with a reduced environmental footprint that are less toxic to natural enemies.

Implications for Sustainable Pest Management

The shift towards conservation biological control reflects a broader recognition of the importance of sustainability in agricultural practices. By fostering environments that naturally control pest populations, this approach not only reduces the need for chemical interventions but also promotes ecological balance and resilience. These practices not only enhance the immediate effectiveness of biological control agents but also contribute to long-term sustainability by preserving the health of the agricultural ecosystem.

3. Use of Microbial Control Agents

The utilization of microbial control agents represents a growing trend in biological control, driven by their specificity, safety, and environmental compatibility. These agents—comprising bacteria, fungi, and viruses—target specific pests while minimizing impacts on non-target species and ecosystems. Recent advancements in formulation and delivery methods are enhancing their efficacy and user-friendliness, making them an increasingly viable option for pest management.

Types of Microbial Control Agents

1. **Bacteria**
    - **Example**: *Bacillus thuringiensis* (Bt), a widely used bacterial control agent, produces toxins that are lethal to specific insects upon ingestion. It is effective against a variety of pests, including caterpillars and beetles, without harming beneficial insects or vertebrates.
2. **Fungi**
    - **Example**: *Beauveria bassiana* and *Metarhizium anisopliae* are entomopathogenic fungi that infect and kill insects by directly penetrating their cuticle. They are particularly valuable in controlling pests like termites, thrips, and other soil-dwelling insects.
3. **Viruses**
    - **Example**: Baculoviruses are specific to insects and have been developed as biopesticides for controlling agricultural pests such as the gypsy moth and codling moth larvae. They cause disease within the host insect that is lethal, yet are harmless to humans, animals, and beneficial insects.

Advancements in Formulation and Delivery

The effectiveness of microbial agents is heavily dependent on how they are formulated and delivered. Recent technological advancements have significantly improved these aspects:

- **Enhanced Formulations**: Microencapsulation and other modern formulation technologies protect microbial agents from environmental degradation and extend their shelf life. This encapsulation can also control the release rate of the active agents, increasing their effectiveness and persistence in the field.
- **Improved Delivery Systems**: Innovations in delivery mechanisms, such as drone technology for aerial applications and advanced spraying equipment, ensure more precise and effective distribution of microbial agents across large areas or difficult terrain.

Safety and Environmental Benefits

One of the most compelling aspects of microbial control agents is their safety profile. They offer several environmental benefits:

- **Specificity**: Their high specificity ensures that they target only the intended pests, reducing the risk of disrupting non-target species, including pollinators and other beneficial insects.
- **Biodegradability**: As naturally occurring organisms, they do not leave harmful residues in the environment, contrasting sharply with chemical pesticides.
- **Resistance Management**: Using microbial agents can help manage resistance in pest populations, especially when integrated with other control methods.

4. Expansion into New Crops and Regions

The expansion of biological control into new crops and regions marks a significant trend in the field of pest management. This trend is largely driven by the global need for sustainable and environmentally friendly approaches to agricultural pest control. As the demand for safer food production increases and new agricultural markets emerge, biological control is being adapted to manage pests in both high-value crops and newly cultivated areas.

Adoption in High-Value Crops

High-value crops, such as fruits, vegetables, and ornamentals, are particularly susceptible to pest damage, which can significantly affect both yield and quality. The use of biological control in these crops offers several advantages:

- **Reduced Chemical Residues**: Biological control reduces the reliance on chemical pesticides, important for crops that are consumed fresh.
- **Market Preferences**: There is a growing consumer preference for products grown with minimal chemical inputs, driving the adoption of biological methods.
- **Regulatory Compliance**: Tighter regulations on pesticide residues in many markets encourage the use of biological control agents as a compliant alternative.

For example, the use of parasitoid wasps to control aphids in greenhouse vegetable production has seen substantial growth, providing effective control without the use of harmful chemicals.

Expansion into New Agricultural Regions

As agricultural practices expand into new regions, often driven by the changing climate or economic opportunities, there is a parallel need to manage new or exacerbated pest challenges. Biological control offers a viable solution by providing:

- **Adaptability**: Biological control agents can often be selected or adapted to suit specific regional conditions, making them effective even in non-native settings.
- **Environmental Conservation**: In regions where agriculture is newly developed, maintaining ecological balance is crucial. Biological control minimizes ecological disruption compared to conventional pesticides.
- **Sustainability**: As new agricultural areas develop, establishing sustainable practices from the outset is preferable to mitigate long-term environmental impacts.

Challenges and Opportunities

The expansion of biological control into new crops and regions also presents unique challenges and opportunities:

- **Biological and Ecological Research**: Understanding the specific ecological dynamics of new regions or crops is essential for the successful implementation of biological control. This might involve detailed studies into local pest and predator species, climate conditions, and crop characteristics.
- **Infrastructure Development**: Establishing the necessary infrastructure to support biological control, such as local production facilities for biological agents or training programs for farmers, is crucial for its success.

- **Policy and Regulatory Frameworks**: Developing supportive policies and regulatory frameworks can accelerate the adoption of biological control. This includes easing the registration processes for biological agents and providing incentives for farmers to adopt sustainable practices.

D. Future Possibilities of Biological Control

The future of biological control holds significant promise, driven by ongoing research and innovation. Key possibilities include:

1. Enhanced Agent Discovery and Development

The future of biological control is vibrant and holds considerable promise for advancing sustainable agriculture. As research continues and technological capabilities expand, significant potential lies in the enhanced discovery and development of new biological control agents. These advancements are pivotal in providing more sophisticated, effective, and targeted tools for pest management.

Enhanced Discovery of New Agents

Ongoing exploration into diverse ecosystems and the microorganisms they harbor can lead to the discovery of novel biological control agents. This exploration is not limited to terrestrial environments but also includes aquatic systems and extreme environments where unique organisms might exist with potential pest control properties. The key to this exploration includes:

- **Bioprospecting**: Systematic search for biological substances in nature that may be developed into commercial products, including new biological control agents.
- **Taxonomic Studies**: Deepening understanding of biodiversity, particularly lesser-known insect species and microorganisms, can reveal new candidates for biological control.

Advances in Biotechnology and Genomics

The application of cutting-edge biotechnological and genomic tools is transforming the development of biological control agents. These technologies offer the following advantages:

- **Precision Engineering**: Techniques like CRISPR-Cas and other gene-editing tools allow for the precise modification of biological control agents to enhance their effectiveness, specificity, and resilience.
- **Rapid Screening and Optimization**: High-throughput genetic screening can quickly identify the most effective strains or species for controlling specific pests.

- **Synthetic Biology**: The design and construction of new biological parts, devices, and systems, or the redesign of existing, natural biological systems for useful purposes, which can include creating entirely new biological control agents that are more efficient and specific.

Targeted Control Agents

With advancements in molecular biology and a deeper understanding of pest and predator interactions, future biological control agents can be designed to target specific pests with unprecedented precision. This specificity minimizes impacts on non-target species and aligns with environmental conservation goals. Innovations might include:

- **Species-specific Pathogens**: Microbial control agents that are pathogenic only to the target pest species.
- **Behavior Modification**: Agents that alter the behavior of pests, reducing their reproductive success or deterring them from damaging crops.

Implications for Pest Management

The enhanced discovery and development of biological control agents have broad implications for pest management, including:

- **Increased Efficacy and Safety**: More effective and safer pest management solutions that reduce the need for chemical pesticides.
- **Sustainability**: Support for sustainable agricultural practices that protect the environment while maintaining high productivity.
- **Economic Benefits**: Reduction in crop losses and decreased reliance on chemical inputs can significantly benefit the agricultural economy.

2. Improved Integration with Digital Agriculture

The integration of biological control with digital agriculture is poised to revolutionize how pest management is implemented in modern farming systems. As digital technologies advance, they offer powerful tools for enhancing the precision, effectiveness, and timeliness of biological control strategies. This integration includes the use of precision farming techniques, decision support systems, and real-time data analytics to optimize biological control interventions.

Precision Farming and Biological Control

Precision farming technologies utilize data from various sources, such as satellite imagery, drones, and sensor networks, to manage agricultural practices precisely and efficiently. When integrated with biological control, these technologies can:

- **Targeted Deployment**: Guide the precise application of biological control agents to the areas most in need, minimizing waste and maximizing impact.
- **Habitat Mapping**: Use geospatial data to identify and enhance habitats favorable to natural enemies, supporting their proliferation and effectiveness.
- **Monitoring Pest Dynamics**: Employ remote sensing and other monitoring technologies to track pest populations and movements, allowing for timely and proactive biological control measures.

Decision Support Systems

Decision support systems (DSS) analyze data and provide insights and recommendations to farmers and land managers. In the context of biological control, these systems can:

- **Optimize Intervention Timing**: Analyze environmental data, pest life cycles, and control agent performance to determine the optimal times for introducing or enhancing biological control measures.
- **Scenario Simulation**: Allow users to simulate various control scenarios and predict their outcomes, helping to make informed decisions about the use of biological control agents.
- **Integrated Pest Management (IPM) Advice**: Combine data on chemical, cultural, and biological control methods to provide holistic and sustainable pest management strategies.

Real-Time Monitoring and Management

The capability to monitor pest and control agent populations in real time provides a significant advantage in managing agricultural pests. This can be achieved through:

- **Sensor Networks**: Deploy sensors to detect and quantify pest populations and environmental conditions affecting both pests and biological control agents.
- **Mobile Technology**: Use apps and mobile devices to collect and transmit real-time data from the field to decision-makers, enabling swift responses to emerging pest issues.
- **Artificial Intelligence (AI) and Machine Learning**: Utilize AI to analyze large datasets from multiple sources, improving the prediction of pest outbreaks and the assessment of control strategies' effectiveness.

Implications for Pest Management

The improved integration of biological control with digital agriculture technologies has several implications:

- **Enhanced Efficiency**: More precise and efficient use of biological control agents reduces costs and environmental impact.
- **Increased Effectiveness**: Real-time data and advanced analytics improve the timing and specificity of biological control interventions, leading to better pest suppression.
- **Sustainability**: By reducing reliance on chemical inputs and enhancing natural pest control mechanisms, this integration supports more sustainable agricultural practices.

3. Climate Change Adaptation

As the global climate continues to change, the challenges presented by shifting pest distributions and new agricultural pests are becoming increasingly prominent. Biological control stands as a critical tool in adapting to these changes, offering sustainable solutions to manage emerging pest threats. The adaptability and effectiveness of biological control agents in the face of climate change will rely heavily on focused research and development aimed at understanding and enhancing their climate resilience.

Adapting to Shifting Pest Distributions

Climate change affects temperature patterns, precipitation, and seasonal cycles, all of which can influence pest behaviors and distributions. As traditional geographic boundaries of pests shift, new areas may become vulnerable to infestations previously unknown to them. In this context, biological control can offer adaptive solutions:

- **Targeted Agent Deployment**: Utilizing biological control agents in areas where new pests emerge as a result of climate shifts ensures targeted and effective pest management without the extensive use of chemical pesticides.
- **Dynamic Monitoring Systems**: Leveraging climate and environmental data to predict pest migrations and outbreaks can help in preemptively deploying biological control agents.

Research into Climate Resilience of Biological Control Agents

Ensuring the continued effectiveness of biological control agents under changing climate conditions requires significant research into their resilience and adaptability:

- **Phenotypic Plasticity and Genetic Diversity**: Studies on the phenotypic plasticity of biological control agents can provide insights into their ability to adapt to various environmental conditions. Understanding and enhancing genetic diversity within agent populations can also help increase their resilience.
- **Environmental Tolerance Ranges**: Researching the tolerance ranges of biological control agents to different temperatures, humidities, and other climatic conditions is vital. This information can guide the selection of agents most likely to succeed under anticipated future climates.
- **Simulating Future Conditions**: Utilizing climatic chambers and simulation models to test the performance of biological control agents under various future climate scenarios can help predict their long-term viability and effectiveness.

Integrating Climate Adaptation Strategies

Integrating biological control into broader climate adaptation strategies for agriculture involves several key components:

- **Policy Support**: Developing policies that support the research and deployment of biological control agents as part of climate adaptation strategies in agriculture.
- **Collaboration Among Stakeholders**: Encouraging collaboration between researchers, farmers, policy-makers, and the community to develop and implement effective biological control strategies that are responsive to climate change.
- **Education and Training**: Providing resources and training to agricultural practitioners on the latest developments in climate-resilient biological control practices.

4. Increased Public Awareness and Support

As biological control continues to evolve as a sustainable alternative to chemical pesticides, increasing public awareness and garnering support becomes crucial for its widespread adoption and success. Effective education and outreach efforts are essential to communicate the benefits of biological control, emphasizing its environmental and economic advantages.

Importance of Public Awareness

Public perception plays a pivotal role in the adoption of innovative agricultural practices. Enhancing public awareness about biological control involves:

- **Highlighting Environmental Benefits**: Educating the public about the reduced environmental impact of biological control compared to traditional chemical methods, including its non-toxic nature and its role in preserving biodiversity.

- **Emphasizing Economic Advantages**: Demonstrating the cost-effectiveness of biological control through reduced pesticide use and enhanced crop yields can appeal to both farmers and consumers.
- **Showcasing Success Stories**: Sharing concrete examples of successful biological control cases helps illustrate its effectiveness and reliability.

Educational Initiatives

To build support and understanding, educational initiatives should target multiple levels of society:

- **Schools and Universities**: Integrating topics on biological control within biology and agriculture curricula to educate the next generation of scientists, farmers, and informed consumers.
- **Public Outreach Campaigns**: Using media, workshops, and public demonstrations to spread knowledge and generate interest among the broader public.
- **Farmer Training Programs**: Providing practical, hands-on training to farmers on implementing biological control in their crop management systems.

Policy Support

Securing support from policymakers is essential for the growth and integration of biological control into mainstream agricultural practices. Policy measures that can promote biological control include:

- **Funding for Research**: Allocating government funds for research on biological control to innovate and improve technologies.
- **Regulatory Frameworks**: Simplifying the process for approval and use of biological control agents to make them more accessible.
- **Incentives for Farmers**: Offering financial incentives for farmers who adopt biological control methods, such as tax breaks or subsidies.

Stakeholder Engagement

Building a coalition of stakeholders including researchers, agricultural professionals, environmental groups, and agribusinesses can help drive the adoption of biological control. This involves:

- **Partnerships Between Academia and Industry**: Collaborating on research projects and the commercialization of biological control products.

- **Community-Based Programs**: Engaging local communities in developing and implementing biological control methods that are tailored to their specific environmental and agricultural conditions.

E. Importation of Natural Enemies: Quarantine Regulations

The importation of natural enemies for biological control involves complex processes governed by stringent quarantine regulations. These regulations are designed to ensure that the introduction of biological control agents does not inadvertently harm native ecosystems, agriculture, or public health.

Overview of Quarantine in Biological Control

Quarantine protocols play a pivotal role in the field of biological control, serving as the first line of defense against potential ecological disruptions that might arise from the introduction of non-native biological control agents. These protocols are meticulously designed to prevent any adverse effects on native species and ecosystems while ensuring compliance with international biodiversity and biosecurity standards.

Primary Objectives of Quarantine

Quarantine regulations in biological control are crafted with several key objectives in mind:

1. **Prevention of Non-target Effects**: This is arguably the most critical objective. The biological control agents are assessed rigorously to ensure that their introduction will not harm indigenous species. This involves detailed ecological studies and host-range testing to confirm that the agents specialize in targeting only the intended pest species.
2. **Avoidance of Pathogen Spread**: Biological control agents, like any living organisms, can carry pathogens. Quarantine measures are essential to ensure that these agents do not introduce new diseases to native wildlife, agriculture, or humans. This is achieved through health screenings and pathogen load assessments conducted under strict laboratory conditions.
3. **Compliance with International Standards**: The global nature of biological control—often involving the movement of biological agents across borders—requires adherence to international agreements such as the Convention on Biological Diversity (CBD) and the International Plant Protection Convention (IPPC). These agreements facilitate cooperation between countries and ensure that biological control practices do not contravene international laws or harm global biodiversity.

The Quarantine Process

The quarantine process for biological control agents is a complex, multi-stage procedure designed to mitigate any risks associated with their introduction. This process is critical for ensuring that the agents are both effective against pests and safe for the ecosystem into which they are introduced.

Pre-Import Risk Assessment

Before any biological agent is imported, a comprehensive risk assessment is conducted to evaluate its potential ecological impact. This assessment includes:

- **Host Specificity Testing**: Detailed tests are conducted to ensure the agent targets only the specified pest. This involves controlled environment testing where the agent is exposed to a variety of potential host species to observe any off-target effects.
- **Environmental Impact Studies**: These studies assess the potential impact of the agent on the local ecosystem. They consider factors such as competition with native species, potential for the agent to become invasive, and any indirect effects on other species, such as through changes in food web dynamics.
- **Pathogen Risk Assessment**: This involves checking the agent for any pathogens it may carry that could affect other organisms in the new environment. Pathogen risk assessments are particularly rigorous, involving both genetic and microbiological analyses to detect any known or unknown pathogens.

Source Country Regulations

The country from which a biological control agent is sourced also plays a crucial role in the quarantine process. These countries often have their own set of regulations designed to prevent the export of harmful organisms:

- **Export Controls and Inspections**: Biological agents are typically subject to detailed inspections before export. These inspections ensure that the agents are free from diseases and are the correct species as per the international permits.
- **Documentation and Traceability**: Exporting countries are required to provide comprehensive documentation that traces the origin of the agents, their breeding conditions, and any treatments they have undergone before export. This documentation is crucial for ensuring transparency and for facilitating smooth passage through quarantine procedures in the importing country.

Import Permits and Documentation

The process of importing biological control agents involves detailed documentation and the acquisition of specific permits:

- **Import Permits**: These are issued by the importing country's agricultural or environmental authority and stipulate the conditions under which the agents must be imported and handled.
- **Health Certificates**: Issued by the exporting country, these certify that the biological control agents are free from pests and diseases and have been handled in accordance with international biosecurity standards.
- **Previous Use Evidence**: This includes data and case studies demonstrating the effectiveness and safety of the agents in other regions. This evidence helps to reassure regulatory bodies that the risk of unforeseen negative impacts is minimal.

Quarantine Facilities

Upon arrival, agents are kept in state-of-the-art quarantine facilities designed to prevent any accidental release into the environment:

- **Containment Measures**: These facilities are equipped with advanced containment measures such as air filters, double-door entry systems, and sterilization procedures for any waste material.
- **Monitoring Protocols**: The agents are monitored regularly for any signs of disease or unexpected behavior. This monitoring is critical for early detection of potential problems.
- **Controlled Testing Environment**: Further testing is often conducted within these facilities to confirm the agents' effectiveness and safety under local environmental conditions.

Post-Release Monitoring

After the biological control agents are released, a rigorous monitoring program is implemented:

- **Population Dynamics Tracking**: This involves regular surveys and population modeling to track the growth and spread of the agent and the target pest populations.
- **Ecosystem Impact Assessments**: Ongoing ecological studies assess the long-term impacts of the biological control agents on the local environment.
- **Adaptive Management Techniques**: Based on the outcomes of monitoring, management strategies may be adjusted to optimize the

effectiveness of the biological control measures and mitigate any negative effects.

Legal and Regulatory Frameworks

The regulatory frameworks that underpin quarantine measures are complex and vary significantly between countries. However, they are all designed to align with international standards to ensure a uniform approach to biosecurity.

- **National Legislation**: Each country develops its own comprehensive biosecurity legislation that governs the importation of biological control agents. These laws are crafted to reflect the specific ecological, agricultural, and social contexts of the country.
- **International Agreements**: Frameworks like the CBD and IPPC provide guidelines and standards that help shape national regulations. These international agreements ensure that while countries have the freedom to develop their own biosecurity measures, they must also cooperate to prevent transboundary ecological impacts.

Challenges and Future Directions

The future of quarantine measures in biological control will need to address several emerging challenges:

- **Global Trade and Climate Change**: As global trade increases and climate patterns shift, new pest pathways and pressures are emerging. Quarantine measures will need to evolve to address these new challenges.
- **Advances in Technology**: Technological advances in genomics and biotechnology are making it possible to engineer biological control agents with specific traits. Quarantine regulations will need to keep pace with these technologies to ensure they are used safely.
- **Public Engagement**: As public awareness of environmental issues increases, there is a growing demand for transparency and safety in the use of biological control agents. Public engagement strategies will be critical for maintaining trust and support for biological control programs.

F. Biotechnology in Biological Control

Biotechnology is revolutionizing the field of biological control, providing innovative solutions for managing agricultural pests and diseases more effectively and sustainably. Through genetic engineering, microbial biotechnology, and RNA interference (RNAi), biotechnology offers precise and targeted approaches to pest control, minimizing environmental impact while enhancing crop protection.

Genetic Engineering and Biological Control Agents

Genetic engineering has opened up new possibilities for enhancing the capabilities of biological control agents. This technology allows for the precise modification of an organism's genetic material to improve its effectiveness against pests.

- **Enhancement of Agent Traits**: Genetic modifications can increase the robustness of biological control agents, allowing them to withstand adverse environmental conditions, such as extreme temperatures or low moisture levels. For example, modifying fungal agents to thrive in dry conditions can significantly extend their applicability and effectiveness in arid regions.
- **Improvement of Host Specificity**: Through genetic engineering, control agents can be made highly specific to their target pests. This reduces the risk of affecting non-target species, a crucial aspect in maintaining ecological balance. For instance, certain genes responsible for host recognition can be enhanced or modified to ensure that the control agent affects only the intended pest species.
- **Safe Deployment Measures**: Genetic modifications often include safety features to prevent unintended spread or persistence in the environment. This can involve the incorporation of genetic 'kill switches' that can deactivate the agent under specific conditions, or confinement systems that restrict the agent to a particular geographic area or set of environmental conditions.

Microbial Biotechnology

Microbial agents such as bacteria, fungi, and viruses provide a biological alternative to chemical pesticides, offering targeted pest control with reduced environmental impact. Advances in microbial biotechnology have significantly improved the efficacy and usability of these agents.

- **Enhanced Virulence and Efficacy**: Microbial agents can be engineered to increase their pathogenicity towards specific pests, ensuring more effective control. For example, certain bacteria can be genetically altered to produce higher quantities of toxins that are lethal to specific insects but harmless to others and to humans.
- **Stabilization and Formulation Improvements**: Innovations in formulation technology help in stabilizing these agents under various environmental conditions, thereby extending their shelf life and maintaining efficacy when applied in the field. Techniques such as encapsulation protect microbial agents from UV degradation and provide controlled release, which enhances their longevity and effectiveness.

- **Precision Delivery Systems**: Advanced delivery technologies, such as drone-based spraying or microencapsulation techniques, allow for the precise application of microbial agents. This ensures that the agents are delivered directly to the area where they are needed, minimizing wastage and reducing the volume of agents required.

RNA Interference (RNAi)

RNA interference is a cutting-edge technology that offers a novel way to control pests by silencing specific genes responsible for their survival, reproduction, or invasiveness.

- **Gene Targeting**: RNAi technology can target specific genes in pests that are crucial for their survival, such as those involved in reproduction or digestion. Silencing these genes can lead to reduced pest populations without the use of chemical insecticides.
- **Adaptability to Pest Resistance**: As pests evolve resistance to traditional pesticides, RNAi provides an adaptable tool to manage such challenges. New RNAi constructs can be rapidly developed to target alternative genes in resistant pest populations, offering a way to stay ahead of resistance development.
- **Integration with Existing IPM Strategies**: RNAi can be integrated seamlessly into existing integrated pest management (IPM) strategies, providing an additional tool that complements chemical, cultural, and other biological control methods. This integration can help reduce overall pesticide use and promote more sustainable pest management practices.

Challenges and Ethical Considerations

While the benefits of biotechnology in biological control are substantial, several challenges and ethical considerations need to be addressed:

- **Complex Regulatory Environment**: The introduction of genetically modified organisms (GMOs) into the environment is heavily regulated. Navigating these regulations requires significant time and resources, potentially slowing the development and deployment of new biological control agents.
- **Public Perception and Acceptance**: GMOs often face public skepticism and resistance based on safety concerns and ethical considerations. Transparent communication and public engagement are essential to build trust and acceptance of biotechnologically derived biological control methods.

- **Biodiversity and Ecological Impacts**: The potential long-term impacts of genetically engineered organisms on biodiversity are a concern. Continuous monitoring and comprehensive ecological assessments are necessary to ensure that these agents do not disrupt local ecosystems or lead to unintended environmental consequences.

G. Semiochemicals in Biological Control

Semiochemicals, a diverse group of chemicals used for communication among organisms, play a crucial role in biological control strategies. These compounds mediate interactions within and between species, influencing behaviors such as mating, foraging, and predator evasion. In the context of pest management, semiochemicals are harnessed to manipulate pest and natural enemy behaviors, enhancing the efficacy of biological control methods. This approach not only targets pests more precisely but also reduces reliance on chemical pesticides, promoting more sustainable agricultural practices.

Types of Semiochemicals Used in Biological Control

Semiochemicals are categorized based on the nature of the interaction they influence. For biological control, the most relevant types are pheromones and allelochemicals:

1. **Pheromones**: These are compounds that organisms use to communicate within the same species. They play a pivotal role in mating, aggregating, or dispersing individuals. Pheromones are extensively used in pest management for:
   - **Mating Disruption**: Synthetic pheromones are released to confuse male pests, preventing them from locating females and thus reducing reproduction rates.
   - **Attraction**: Pheromones can attract pests to traps or areas treated with biological agents or pesticides, concentrating the pest population and enhancing control efficiency.
2. **Allelochemicals**: These are chemicals that mediate interactions between different species and include allomones, kairomones, and synomones:
   - **Allomones**: Beneficial to the emitter, allomones can repel or deter pests. For example, certain plants emit allomones that make them less palatable or attractive to herbivores.
   - **Kairomones**: These are advantageous to the receiver but not the emitter. Predators and parasitoids often use kairomones emitted by their prey or host to track them down.

- **Synomones**: These chemicals benefit both the emitter and the receiver, such as when plants release volatile organic compounds that attract natural enemies of their herbivores.

Applications of Semiochemicals in Pest Management

The use of semiochemicals in pest management involves several innovative applications that enhance the control of pest populations while minimizing environmental impacts:

- **Trap Cropping**: Plants that emit strong synomones are used as trap crops to attract natural enemies, effectively creating a biological control hotspot where pests are actively managed by their natural predators.
- **Push-Pull Strategies**: This strategy combines the use of repellent plants (push) and attractive trap plants (pull) to manage pest distribution. Pests are 'pushed' away from the main crop by repellent allelochemicals and 'pulled' towards an area where they can be more easily controlled or eliminated.
- **Integrated Pest Management (IPM)**: Semiochemicals are integrated with other biological control methods, cultural practices, and chemical controls in an IPM approach. This integration helps to optimize pest control by exploiting multiple avenues of pest management, thereby reducing the chance of pest resistance and enhancing long-term sustainability.

Benefits and Challenges of Using Semiochemicals

Benefits

- **Target Specificity**: Semiochemicals offer high specificity, typically affecting only targeted pest species or closely related groups. This specificity helps to preserve beneficial insects and non-target organisms.
- **Environmental Safety**: Unlike broad-spectrum chemical pesticides, semiochemicals do not persist in the environment and degrade quickly, reducing environmental contamination and the risk of non-target effects.
- **Resistance Management**: Using semiochemicals as part of a broader IPM strategy can help mitigate the development of resistance to conventional pesticides.

Challenges

- **Complexity of Semiochemical Interactions**: The effectiveness of semiochemicals can be influenced by environmental factors such as temperature, humidity, and the presence of other chemical signals, which can complicate their use in field conditions.

- **Cost and Scalability**: The synthesis and deployment of semiochemicals can be expensive, and scaling up their use for large agricultural applications can be economically challenging.
- **Regulatory and Approval Processes**: As with any pest control product, semiochemicals must undergo rigorous regulatory review to ensure they are safe and effective, which can be a lengthy and costly process.

Multiple-choice Questions (MCQs)

1. What is the primary objective of quarantine regulations in biological control?

   A) Increase the efficiency of pesticides

   B) Prevent non-target effects and pathogen spread

   C) Enhance the fertility of soil

   D) Promote the use of chemical controls

   Answer: B

2. Which biotechnological tool is used to silence specific genes in pest management?

   A) DNA sequencing B) RNA interference (RNAi)

   C) Polymerase chain reaction (PCR) D) Gel electrophoresis

   Answer: B

3. What role do pheromones play in biological control?

   A) Disrupt mating of pests B) Increase crop yields

   C) Act as a herbicide D) Improve soil quality

   Answer: A

4. Which of the following is a benefit of using semiochemicals in pest management?

   A) Long environmental persistence B) High specificity and safety

   C) Induces pest resistance D) Broad-spectrum pest control

   Answer: B

5. How are genetically modified organisms (GMOs) regulated in the context of biological control?

   A) They are unregulated

   B) Only in developing countries

   C) Through international biosecurity standards

D) Based on the local community rules

Answer: C

6. What is the primary focus of conservation biological control?

   A) Use of chemical pesticides

   B) Modification of the environment to support natural enemies

   C) Breeding plants to be more pest-resistant

   D) Introduction of non-native pest species

   Answer: B

7. Which technique involves the use of trap crops in biological control?

   A) Push-pull strategy B) Chemical control

   C) Crop rotation D) Genetic engineering

   Answer: A

8. What is the primary function of kairomones in biological control?

   A) Benefit the emitter by deterring pests

   B) Benefit the receiver by locating food or hosts

   C) Increase crop resistance to diseases

   D) Repel natural enemies

   Answer: B

9. What is the role of microbial agents in biological control?

   A) Enhance the growth of pest populations

   B) Decompose organic matter in soil

   C) Target and control pest populations

   D) Increase nitrogen fixation in plants

   Answer: C

10. How does RNA interference contribute to pest management?

    A) By enhancing plant growth

    B) By silencing specific pest genes

    C) By increasing genetic diversity in crops

D) By promoting beneficial bacterial growth in soil

Answer: B

11. What is the purpose of DNA barcoding in biological control?

A) To enhance plant nutrition

B) To identify and track biological control agents

C) To increase water absorption in plants

D) To monitor soil pH levels

Answer: B

12. Why are semiochemicals considered environmentally friendly in pest management?

A) They persist long in the environment

B) They are non-specific to target species

C) They naturally degrade and have minimal non-target effects

D) They enhance soil mineral content

Answer: C

13. What advantage does genetic engineering offer in developing biological control agents?

A) Decreased efficacy

B) Inability to target specific pests

C) Improved host specificity and effectiveness

D) Reduction in crop yields

Answer: C

14. How does the importation of natural enemies involve quarantine regulations?

A) To ensure they are free from diseases

B) To increase their reproductive rates

C) To reduce their effectiveness

D) To enhance their genetic diversity

Answer: A

15. Which international agreement influences quarantine measures in biological control?

A) Kyoto Protocol

B) Geneva Conventions

C) Convention on Biological Diversity (CBD)

D) Treaty of Versailles

Answer: C

16. What does the push-pull strategy in biological control involve?
   A) Using only chemical controls
   B) Employing both attractive and repellent plants
   C) Pushing pests towards the crop
   D) Pulling natural enemies away from the crop

   Answer: B

17. Why is ongoing monitoring important after releasing a biological control agent?
   A) To ensure the agent does not become invasive
   B) To reduce the effectiveness of the agent
   C) To decrease biodiversity
   D) To enhance the use of pesticides

   Answer: A

18. What is a major challenge in using semiochemicals for pest control?
   A) They are too specific
   B) They quickly degrade in the environment
   C) They increase the pest resistance
   D) They are readily accepted by the public

   Answer: B

19. How does habitat manipulation work in conservation biological control?
   A) By decreasing the diversity of natural enemies
   B) By modifying the environment to favor natural enemies
   C) By introducing genetically modified crops
   D) By applying chemical fertilizers

   Answer: B

20. What is the role of microbial formulation technology in biological control?
   A) Decreases the stability of microbial agents
   B) Enhances the effectiveness and stability of microbial agents
   C) Prevents microbial agents from acting on pests

D) Reduces the specificity of microbial agents

Answer: B

21. Which factor complicates the field use of semiochemicals in biological control?

A) Predictable effects
B) Low cost
C) Environmental variability
D) High stability

Answer: C

22. What future trend is expected to influence the development of biological control agents?

A) Decreased emphasis on environmental sustainability
B) Increased regulatory restrictions without scientific basis
C) Advancements in biotechnology and genomics
D) Shift towards exclusive use of chemical pesticides

Answer: C

23. How are allelochemicals different from pheromones in their use in biological control?

A) Allelochemicals are not used in biological control
B) Allelochemicals mediate interactions between different species
C) Pheromones are harmful to the environment
D) Pheromones increase biodiversity

Answer: B

24. What is a primary benefit of integrating semiochemicals with other IPM strategies?

A) It simplifies pest management practices
B) It enhances the effectiveness of overall pest management
C) It encourages the use of more chemical pesticides
D) It reduces the need for monitoring pest populations

Answer: B

25. Why is public perception a challenge for the adoption of genetically modified biological control agents?

A) Lack of understanding and potential safety concerns
B) Universal acceptance of genetically modified organisms

C) No regulatory challenges associated with GMOs

D) GMOs are less effective than traditional methods

Answer: A

Short Questions

1. What is the primary function of pheromones in biological control?
2. Name a biotechnological method used to enhance the effectiveness of biological control agents.
3. Which type of semiochemical benefits both the emitter and the receiver?
4. What does RNA interference target in pest management?
5. List one benefit of using microbial agents in biological control.

Long Questions

1. Explain the role of quarantine regulations in the importation of natural enemies for biological control and discuss how these regulations prevent ecological disruptions.
2. Describe how genetic engineering and microbial biotechnology are used to improve the specificity and effectiveness of biological control agents. Include examples of how these technologies are applied in the field.
3. Discuss the concept of the push-pull strategy in biological control. How does integrating semiochemicals with other pest management strategies enhance the overall efficacy of this approach?